The Rhetoric of Science in the Evolution of American Ornithological Discourse

ATTW Contemporary Studies in Technical Communication

M. Jimmie Killingsworth, Series Editor

Published in cooperation with the Association of Teachers of Technical Writing

Volume 1:
Foundations for Teaching Technical Communication: Theory, Practice, and Program Design
edited by Katherine Staples and Cezar M. Ornatowski, 1997

Volume 2:
Writing at Good Hope: A Study of Negotiated Composition in a Community of Nurses
by Jennie Dautermann, 1997

Volume 3:
Computers and Technical Communication: Pedagogical and Programmatic Perspectives
edited by Stuart A. Selber, 1997

Volume 4:
The Practice of Technical and Scientific Communication: Writing in Professional Contexts
edited by Jean A. Lutz and C. Gilbert Storms, 1998

Volume 5:
The Dynamics of Writing Review: Opportunities for Growth and Change in the Workplace
by Susan M. Katz, 1998

Volume 6:
Essays in the Study of Scientific Discourse: Methods Practice, and Pedagogy
edited by John T. Battalio, 1998

Volume 7:
Three Keys to the Past: The History of Technical Communication
edited by Teresa C. Kynell and Michael G. Moran, 1998

Volume 8:
The Rhetoric of Science in the Evolution of American Ornithological Discourse
by John T. Battalio, 1998

The Rhetoric of Science in the Evolution of American Ornithological Discourse

by
John T. Battalio
Boise State University

Volume 8 in ATTW Contemporary Studies in Technical Communication

Ablex Publishing Corporation
Stamford, Connecticut
London, England

Printed in the United States of America

Library of Congress Cataloging-in-Publication Data

Battalio, John T.
The rhetoric of science in the evolution of American ornithological discourse / by John T. Battalio.
p. cm. -- (ATTW contemporary studies in technical communications ; v. 8)
Includes bibliographical references and index.
ISBN 1-56750-395-0 (cloth). -- ISBN 1-56750-396-9 (pbk.)
1. Ornithology-- United States--Authorship--History.
2. Communication in science--United States--History. I. Title.
II. Series.
QL672.6.B38 1998
598'.0973--dc21 98-7695
CIP

Ablex Publishing Corporation
100 Prospect Street
Stamford, Connecticut 06901-1640

Published in the U.K. and Europe by:
Ablex Publishing Corp. (UK)
38 Tavistock Street
Covent Garden
London WC2E 7PB
England

To my parents, Joanne Downs
and Vincent John Battalio.
In loving memory.

Contents

Acknowledgments

I thank the following journals and publishers for granting permission to use the items mentioned below:

- American Ornithologists' Union for permission to reprint seven graphics and three text pages from *The Auk*
- Baywood Publishing Company for permission to reprint excerpts from the following article:
  Battalio, J. T. (1996). The interplay between narrative, education, and exposition in an emerging science. *Journal of Technical Writing and Communication, 26,* 177–191. Copyright © 1996, Baywood Publishing Co., Inc.
- Southern Illinois University Press for permission to reprint and adapt the "Continuum of Scientific Perspectives on Nature" from the following book:
  Killingsworth, M. J., & Palmer, J. S. (1992). *Ecospeak: Rhetoric and environmental politics in America*. Carbondale, IL: Southern Illinois University Press. Copyright © 1992 by the Board of Trustees, Southern Illinois University

I also thank the Ablex staff for their kind assistance, especially Kim Burgos, Acquisitions Editor, and Anne Trowbridge, Book Production Manager and Lisa Barnett, Book Production Editor.

I especially thank those who read portions or all of this manuscript and offered valuable suggestions and criticisms. Having evolved from a doctoral dissertation at Texas A&M University, this book has benefited from the careful guidance of the following committee members: Robert Benson, Barbara lowing professors who kindly agreed to read individual chapters of the manuscript: Jerome A. Jackson, Mike Markel, and Charles Alan Taylor. In particular, I thank Jimmie Killingsworth, ATTW Series Editor, for his continual guidance and encouragement and for his many careful readings and instructive comments throughout the life of this manuscript. Though the book is clearly better due to their suggestions, any failings and shortcomings which remain are mine.

This book represents the merger of a wide variety of interests, which were nurtured by those to whom I owe the most thanks—my parents. I regret that they are no longer here to read these expressions of gratitude. Both of them encouraged me to be independent and to follow the paths I chose for myself, for which I am most grateful. And ultimately the seed for this book was planted by a mother who was so willing to chauffeur a little boy around in search of the birds!

Introduction

Studying science writers as distinct social entities reveals many discourse communities, both among and within specific scientific disciplines, each employing its own characteristic writing processes and styles. Ornithologists constitute one such community of science writers, and it is the American branch of this scientific field which is the focus of this rhetorical study. With more than a century of publication behind them, American ornithologists have left their mark on the history of scientific discourse. In addition to major journals, such as *The Auk*, *Wilson's Bulletin*, and *Condor*, all published continuously since the late 1870s, continued interest in the field has spawned twentieth-century additions, including *American Birds* (now *Field Notes*) and the *Northeastern Bird Bander* (now the *Journal of Field Ornithology*). Other outlets for disseminating ornithology have included state journals and innumerable, still uncataloged, amateur publications.

This study locates ornithological writing in the continuum of scientific discourse and, in the process, reveals the struggle of a scientific discipline on the way to professionalization by describing two related phenomena in the history of science: the influence of amateur discourse on ornithology and the historical development—and conflict—of the perspectives now known as natural history and experimental science. Amateur contributions have been a part of the history of American ornithology since its inception as a scientific field of study in the late nineteenth century. American ornithology, in fact, has been more open than many other scientific disciplines to contributions by amateurs. However, advances in statistical and biological knowledge during the twentieth century have largely closed the field to significant amateur contributions, except in the areas of censusing and field identification. The chances are slim, for instance, that a significant contribution to present-day ornithology will be made by a nonprofessional such as Margaret Morse Nice, whose life history of the song sparrow and bibliographical work are still being referenced some 50 years later.

The controversy over the acceptance of amateur contributions in early American ornithology lends insight into the processes involved in the shaping of professional ornithological discourse. The attitude of Lynds Jones (1905), the editor of the *Wilson Bulletin*, a popular magazine of ornithology, was perhaps typical. In a series of questions to guide readers' field work, he exhorted readers to prove "that the slur often aimed at amateur field work is not applicable in your case at least" (p. 22). In the same year, an editorial in a Western ornithological magazine, the *Condor*, responded to a comment in *Bird-Lore* in which Ernest Thompson Seton suggested that only museum "experts," who best understand the value of birds, be allowed to collect specimens. Editor Walter K. Fisher (1905) chided ornithologists to be tolerant of the beginner and reminded them of how much ornithological knowledge is contributed "by persons who have no connection whatever with museums or institutions of any sort" (pp. 31–32). The diatribe by Althea Sherman (1930) against inaccurate amateur discourse led the March 1930 *Wilson Bulletin*. Dividing ornithology into old and new schools, she derided newcomers who, with only passing interests in and often fallacious knowledge about birds, proceeded to publish ornithologically inaccurate poems, stories, and natural history accounts in popular magazines. She took heart, however, that such "ignorant sentimentalists" (p. 6) had yet to taint the bird magazines. On the professional front, this conflict is most clearly evidenced by J. A. Allen's success in purging the technical literature of the late nineteenth century of descriptions and taxonomic nomenclature based upon botanist-ornithologist-travel writer William Bartram, whose religious upbringing fueled the glorification of God and nature in his writings. Nineteenth-century ornithologists' desire for acceptance by continental zoologists caused them to distance themselves from Bartram's emotionalism, which signified a lack of scientific objectivity.

In general, science began professionalizing during the nineteenth century, making a permanent break with amateur science by the end of the century (see, for instance, Larson, 1977). For example, between 1840 and 1880 British science had effectively professionalized after its break with organized religion. Frank Turner (1978, pp. 362–364) cited a number of significant events: (a) in 1847, the Royal Society reformed its rules to favor more limited membership based on scientific rather than social achievement; (b) by 1870, young scientists who were positivist epistemologists from outside both the inner circle of Victorian elite and the English university system had taken over many positions as editors, professors, and officers in the major societies; and (c) by 1880, there had been a significant rise in the number of academic positions in physics and chemistry.

In the United States, professionalization began to occur after the War of 1812. George Daniels (1967) described the period between 1820 and 1860 as the emergent period of American professional science. American science became esoteric by locating its argumentation in more complex bodies of knowledge. It also began institutionalizing and consciously arguing on behalf of its own legitimation. By the 1830s, positions for scientists in academia and government agencies were increasing (Holmfeld, 1970, p. 22). By 1840, scientific associations, such as those for geology and astronomy, were formed. Of particular significance was the formation of the American Association for the Advancement of Science in 1847. Holmfeld has detailed a controversy over the publication of the proceedings of the 1853 meeting of the Association, a controversy which pitted amateur against professional and which perhaps symbolizes the permanent separation of the two. Soon afterward, the peer review process was initiated, and the Association began publishing its own proceedings, taking control away from local, often amateur, scientists. To further separate themselves from the amateur, American scientists convinced the federal government to found in 1863 a second organization, the National Academy of Sciences, whose members were almost exclusively professional scientists.

However, the professionalization of American ornithology, which I trace in this present study, has been slower, partly because of the continued influence of painter-naturalists such as Audubon and Wilson and the enthusiastic persistence of "bird lovers," whose interest has had both scientific and transcendent components. Still, increasing professionalization, including technological advances and a substantial growth of published data in the field, eventually consolidated the American ornithological discourse community by the mid-twentieth century, further widening the rift between natural history and experimental science perspectives.

The main objective of this study is to trace the rhetorical development of American ornithological discourse, which in turn will show the continual interplay between natural history and experimental science. As its history shows, American ornithology has arisen out of the interest and enthusiasm of both amateur and professional ornithologists, an association that has continued in some form or other since the field's inception during the late nineteenth century. Its discourse then creates a site for study in which amateur scientists mingle with professionals for an extended period of time. Such a study offers a unique opportunity to analyze further the relationship between amateur and expert writing—that is, the relationship between natural history and experimental science discourse—and a means to explore the professionalization of a science.

When I speak of rhetorical development, I am using the term "rhetoric" in its broadest sense. Following classical rhetoric, I seek a better understanding of three influences upon American ornithological discourse: (a) the epistemological and ontological argumentation that formed present-day American ornithology; (b) the forces that molded the unique character of amateur and professional ornithologist; and (c) the interrelationship between a commonly felt love for birds and a pride in one's professional self-identity. Since discourse is situationally placed, I also seek an understanding of the social and historical forces that influenced the argumentation of American ornithologists, as well as the linguistic influences that helped to shape American ornithologists' discourse.

This study of American ornithological discourse is directed to a variety of audiences. It offers scholars in rhetoric and researchers in discourse studies a detailed overview of the processes at work in an evolving scientific discipline. It offers to scientists insights into the relation between scientific writing and the development of scientific communities. It offers historians of science an extended analysis of a professional field, the writings of which have not previously been studied in depth. Finally, it offers the general public, particularly bird enthusiasts, a greater understanding of the origin and evolution of popular interest in birds and a closer look at those who have made the study of birds their livelihood. The ultimate aim is to improve communication by improving our understanding of the social and historical forces that affect communication.

After an overview in chapter 1 of the recent rhetorical, historical, linguistic, and discursive research relevant to this present study, I provide in chapter 2 a historical study of the emergence of American ornithological discourse between 1600 and 1900 in order to determine the major factors influencing the emergence of this discourse. Popular discourse about birds evolved from craft literature, such as falconry manuals, to bird listing, to the natural-history travel narrative. Technical discourse in ornithology, on the other hand, evolved into manuals, field guides, and handbooks resulting from attempts to systematize the study of birds by determining their taxonomic relationships. These two lines of development—popular and technical writing about birds—evolved gradually between 1600 and 1900 into what would become natural history and experimental science discourse, respectively, in American ornithology.

I then propose in chapter 3 discourse models for natural history and experimental science, the beginnings of which in American ornithology were just described. The purpose of these models is to generate a set of characteristic features of discourse to be used as a basis for studying American ornithologists' writing in depth. The models are based on two aspects of discourse: (a) general concepts guiding all writers of discourse, and (b) elements specific to each individual writer. First, rhetorical considerations of persona, audience, and context essential to any discourse model evolve from the answers to the following five questions: (a)Who is the author and for whom is s/he speaking?; (b) What forces bind members together?; (c)

What is their world view?; (d) What is the status of their knowledge and how are knowledge claims made?; and (e) How is the same knowledge used by both groups? Answers to these questions in light of the natural history and experimental science discourse models reveal the following features: the individualism of the naturalist vs. the communal sense of the scientist; the inclusiveness of natural history vs. the exclusiveness of science; a macrocosmic view of "nature " vs. a microcosmic view of specific research goals; and personal vs. professional motivations.

Second, since individual writers make decisions based upon their own past experiences and by the objective norms established by the discourse itself and by audience, a discourse model should also analyze the major linguistic features of individual examples of discourse by representative users of the model discourse. Thus, I use case studies to examine the writings of four ornithological writers. Specifically, the writings of nineteenth-century writer-naturalists Henry David Thoreau and John Burroughs, whose field studies and writing often focused on birds, illustrate natural history writing. In contrast, the writings of two systematic ornithologists, whose DNA research published in *The Auk* has necessitated that their discourse be directed toward the hard-core professional, represent the transformation to professionalization by means of experimental science discourse.

I next show in chapter 4, "Diachronic Changes: The Discourse of Transition," that the two models of discourse described in chapter 3 are not sufficient to describe the full range of ornithologists' writings. In effect, there was not a simple switch by American ornithologists from one discourse to the other, but a gradual changeover from one to the other as American ornithology professionalized. Thus, a third discourse model, that of transitional discourse, is also needed.

I first survey 11 volumes of *The Auk* in 10-year intervals beginning in 1890 in order to determine the changes in topics and focus of American ornithological writing. The survey shows that two lines of development—(a) popular or natural history discourse, and (b) technical or experimental science discourse—have continued to intermingle throughout most of the history of American ornithology during the past one hundred years. Evident are two major shifts in emphasis: one in the early twentieth century when the amount of natural history discourse rises significantly, and the other in the 1970s when experimental science discourse establishes a strong foothold. Before this time, natural history discourse persists in American ornithology, despite the field's increasing professionalization.

Given this gradual intermingling of the two discourses, I use another case study to describe a transitional discourse model—that is, to experimental science. I examine the writings of Frank M. Chapman, early-twentieth-century Curator of Ornithology at the American Museum of Natural History, who wrote for both professionals and lay readers alike. After describing Chapman's selected writings, I compare the linguistic features of his writings with those of the natural history and experimental science discourse models. The analysis suggests the following features as early evidence of the shift to experimental science discourse:

- increases in sentence length, nominalizations, and subject-verb interruptions
- less emphasis on subjects as place holders, the use of the verb *to be*, and the number of simple sentences
- the appearance of both technical and polynomial technical terms
- an emphasis on the reliability of evidence

Suggested also are the last features to establish themselves in experimental science discourse:

- increased use of metadiscourse, definitions, passive voice, and citations
- increased use of simple sentences and a corresponding decrease in the use of most complex clauses
- restricted use of evidentials, particularly expectations
- minimal use of personal pronouns

Chapter 5, "A Hundred Years of American Ornithological Discourse," is a historical study, which explores in greater depth the writings of American ornithologists during their hundred years of professional publication in order to compare their writing with the discourse models previously described. Tracing the evolution of research topics and methodology as published in *The Auk* and interspersing the discussion with examples of a changing American ornithological discourse, I propose four periods through which this discourse has progressed:

PERIOD 1 (1884–1900): Natural History and Experimental Science Discourse in Balance
PERIOD 2 (1900–1940): The Predominance of Natural History Discourse
PERIOD 3 (1940–1970): In Interplay of Discourse
PERIOD 4 (1970–1990): The Emergence of Experimental Science Discourse

I also trace the organization and formatting of ornithological discourse, as well as ornithologists' decisions for displaying their data. After an increase in article length during the first half of the century, *The Auk* was publishing shorter articles, though more of them, by the end of the twentieth century. As further evidence of professionalization, an Introduction,

Methods, Results, and Discussion (IMRaD) format appeared by the end of the 1980 volume, reflecting the presence of a newly standardized discourse. Demonstrating the continuing sophistication of American ornithological discourse and its more exclusively professional audience, data display became more varied and complex through the changing periods. Formal tables registered the largest increases in usage, though line and bar graphs and scatterplots were also increasingly popular choices. Lists, which had been the most popular means of displaying field data early in *Auk* history, had disappeared by the emergence of experimental science discourse. In all, the number of graphics had registered a four-fold increase by the 1990 volume.

In order to analyze a larger sampling of discourse over an extended period of time, I describe in the second half of chapter 5 the results of a second study of articles from *The Auk*. The resulting sample is composed of articles by 44 authors—four per each 10-year period from 1885 through 1990—whose writing I analyze by studying the same 32 linguistic features and using the same methodology as in chapters 3 and 4.

The chapter graphically displays eight linguistic features showing distinct variation: two dealing with sentence complexity, two influencing vocabulary usage, and four influencing the making of factual claims. Serving as the best benchmarks for experimental science discourse, and thus professionalization, are significant increases in the percentage of subject-verb interruptions, nominalizations and technical and polynomial terms, markers for reliability of evidence, and the use of citations. Despite the increasing theoretical complexity and desire for scientific objectivity, four features remained relatively stable: personal pronouns, subordinate causal and conditional clauses, and hedging devices. This stability perhaps reflects the prolonged interrelationship between natural history and experimental science discourse.

Finally, in chapter 6, "Scientific Discourse From the Perspective of American Ornithology," I bring together the historical, social, and linguistic forces impacting American ornithology and end by integrating the results of this analysis of American ornithologists' discourse with studies of other scientific discourse in order to draw more general conclusions about the rhetoric of science. Since a number of generalizations about scientific discourse implicit from previous studies do not hold for American ornithology, I offer a broader perspective of the characteristics of scientific discourse and propose a discourse schema which shows the processes that create a professional scientific discourse.

During the earliest period of American ornithological discourse (1884–1900), a major function of *The Auk* was to describe what it meant to be an American ornithologist, though it was difficult for technical ornithologists to separate themselves from the amateur because much early research was observation-based. In the second period (1900–1940), increased interest in behavioral studies and the rise of conservation and preservation movements altered the research activities of American ornithology. The increasing specialization in *The Auk* during the third discourse period (1940–1970) was influenced by the spread of graduate programs in ornithology and the environmental movement of the 1950s and 1960s, both of which likewise influenced membership in the American Ornithologists' Union.

Based on results from the study's linguistic analyses, I propose 15 linguistic features as *discourse indices* signaling the professionalization of a science—features that should be valuable indicators for other rhetorical and linguistic researchers studying the discourse of emerging sciences. In particular, six features have the greatest predictive value as indices of professionalization because their changes were the most rapid and clearly evident: increasing use of nominalizations, subject-verb interruptions, citations, technical terminology, simple sentences, and high-density graphics like tables and graphs.

Unlike astronomers' and physicists' writing, American ornithological discourse steadily decreased in length after 1960. Unlike astronomers' journals, the volume of publication in *The Auk* declined until the last two decades. Physicists eventually dispersed citations throughout their articles, discussing them more thoroughly. However, the appearance and function of ornithological citations did not change. Physicists' discourse registered no significant change in sentence length or complexity, but ornithologists' writing eventually registered noticeable decreases in sentence length and increases in simple sentences. Although physicists did change the kinds of clauses they used, in almost all cases, clausal usage by ornithologists changed very little. Their writing did show increased complexity, as for instance in the rise of subject-verb interruptions and sentence lengths. Both physicists and ornithologists increased usage of sophisticated graphical presentations. However, though physicists replaced tables with graphs, ornithologists continued to rely on tables the most. In addition, bar graphs eventually superseded all other forms of graphs in *The Auk* as ornithology professionalized. Formatting in physicists' discourse changed from a focus on methods sections to increasing use of discussion, conclusion, and theory sections. Unlike physicists' discourse, methodology sections increased through much of *The Auk*'s history. In addition, though the Acknowledgments section was the last to appear in physicists' discourse, in *The Auk* it was the first section to become standardized. And the complete IMRaD format was not established until 1980, 20 years after the *Physical Review* established its own format. Finally, Bazerman (1988) found a progression in physicists' argumentation from methodology to theory- and problem-oriented articles to model building.

Argumentation in ornithology, however, was more complex because it has been uniformly problem-oriented throughout its history, with methodological argumentation increasing as the science matured; model-building and theory-oriented argumentation have registered only recent births.

Consequently, a study of discourse already in progress does not show the effects of the relationship between evolving discourse and struggles within a scientific field attempting to identify itself as a profession. For *The Auk*, however, American ornithology had not existed previously. Its inception was announced by the appearance of *The Auk*; and its very existence was due to a merger of both amateur and "technical " ornithologist, who formed a community of shared ideas, values, and methodology, which enabled them to call themselves "American Ornithologists' Union " for the first time. There is, then, in the study of American ornithologists' discourse an opportunity to see the changes created as a science develops a sense of its own uniqueness as a profession and begins to cement the bonds that establish itself as a unique, paradigm-bounded community of scientists, whose discourse reflects their sense of professionalism. Thus, more accurate assertions about the processes at work in a developing scientific discourse can be made through a study of an evolving discourse coupled with a jointly evolving professional persona, like that seen in *The Auk* and American ornithology, rather than in a study of discourse from already established professions.

—John T. Battalio
Boise State University
25 December 1997

1

Recent Studies Relevant to an Analysis of American Ornithological Discourse

Though there has been to date no formal rhetorical study of American ornithological discourse, a number of recent studies in areas such as specific scientific disciplines, rhetoric, professionalization, discourse, and linguistics suggest directions for a comprehensive analysis of American ornithologists' writing. The overview below summarizes these studies and provides a preview of the research and the range of topics referenced in this study of American ornithological discourse. The survey is also meant to be an introduction, particularly for those readers unfamiliar with recent work in, or related to, the analysis of scientific discourse. It is not meant to be a comprehensive survey of scholarship in scientific discourse. For a more comprehensive overview of recent scholarship, see Battalio (1998).

STUDIES WITHIN SPECIFIC SCIENTIFIC DISCIPLINES

Especially suggestive for an analysis of American ornithologists' discourse are studies carried out within specific disciplines of modern scientific writing. One study of astronomers' and two of physicists' writing used specific, quantifiable features of these disciplines' discourse in order to study the chronological progression of these fields' discourse. This study of ornithologists' discourse has been designed to parallel the work done by these researchers in order to make direct comparisons among astronomers', physicists', and ornithologists' writing.

Helmut Abt (1981) analyzed 70 years of astronomers' discourse in three American astronomical journals from 1910 to 1980. His study found a tenfold increase in the number of papers and a thirty-fold increase in their length, with nearly a total disappearance of very short papers (that is, those of less than one page) by 1980. These increases occurred primarily from 1940 to 1980. The study concluded that by 1980 authors wrote fewer but longer papers, whose topics included more theoretical issues, a 28% increase by Abt's research. Though there had been a decrease in single-author papers, multiauthored papers (that is, more than five authors) were still rare in 1980.

Two studies have been made of physicists' discourse. A Physics Survey Committee (1973) of the National Academy of Science's National Research Council used bibliographical entries from *Physics Abstracts* and a random sample of American Physical Society members to conclude, like Abt (1981), that papers were becoming longer and co-authorship increasing. However, unlike Abt's analysis of astronomers' discourse, the committee found that physicists wrote more papers per year. The committee's study, however, was limited to scattered 1-year analyses; for instance, only comparisons of 1955 and 1970 membership were used to evaluate changes in frequency of publication. The report clearly showed experimental science's trends toward elitism and separation as described by Latour (1987) and by Killingsworth and Palmer (1992, pp. 108–31). Though the report addressed the need to communicate physics' advancements to the general public, journalists and science writers were left with the task. The committee's comparison of 1958–1959 and 1968 popular literature indices indicated, however, that science writers were not transmitting much physics research to the general public (pp. 1436–1437).

A study by Charles Bazerman (1988) of the *Physical Review*, 1893–1980, was a more comprehensive study of physicists' discourse and has been particularly valuable in this study of American ornithologists' writing. He measured article length; carried out a sentence-level analysis of sentence length, syntax, and word choice; and provided contextual analysis of citations, graphics, organization, and argumentation. Like Abt (1981) and the Physics Survey Committee (1973), Bazerman confirmed the increasing length of science articles during the twentieth century. Through his study of spectroscopic articles, he concluded that this increasing length corresponded to "increasing knowledge and theoretical elaboration" (p. 164), which requires more background and contextual understanding. As a result, use of citations not only increased, but references became dispersed throughout articles and were discussed more thoroughly.

His sentence-level analysis found no significant change in sentence length, complexity, or use of expansion devices. The types of subordinate clauses did change, however, with relative clauses decreasing in number and a corresponding increase in noun clauses and temporal/causal subordinate clauses. He explained this trend by citing increased intellectual complexity. Changes in word choice during the century also supported increased complexity: an increase in abstract over concrete subjects, in substantive active verbs over reportorial ones, and in the number of both technical words and multi-word noun phrases. Graphics likewise became more abstract, with more technically sophisticated graphs replacing tables, as well as a greater use of equations.

Concerning organization, methods sections predominated early in the century, giving way eventually to discussion, conclusion, and theory sections. Section headings were established features after 1950, though Acknowledgments sections were not common until 1960. Bazerman (1988) pointed to increasing use of discussion and conclusion sections after 1930 to infer more issue-oriented discourse. Argumentation during the century evolved from explanation of methodology to theory- and problem-oriented articles in which, after 1930, physicists had to locate their theories within quantum mechanics. In the 1950s and 1960s, theory testing gave way to a modeling approach, indicating a move from the epistemological certainty of theories to more tentative assertions of the appropriateness of one model over another.

RHETORIC OF SCIENCE

Since the appearance of Philip Wander's (1976) definition of the rhetoric of science, there have been innumerable studies showing the presence of rhetoric, or a conscious use of persuasion, in scientific discourse, so much so that most discourse scholars—and scientists—acknowledge the influence of rhetoric in science, at least to some extent. In particular, three recent studies of the rhetoric of science show the value of rhetoric in scientific discourse. They are complementary in that, taken together, they describe well the historical, classical rhetorical, and dialectic components needed in a comprehensive analysis of scientific discourse.

Alan Gross (1990), in *The Rhetoric of Science*, argued that the principles of scientific rhetoric are traceable to principles espoused by classical rhetoric. Present, for instance, in scientific discourse are stasis theory and Aristotelian means of persuasion. Following classical tradition, Gross showed science's concerted use of metaphor by employing tropes and analogies. Linguistic change accompanies science's increasing complexity: Observations and experiments become scientific terms, which in turn become more complex noun phrases as scientific knowledge transforms itself into theoretical networks. Finally, the scientific IMRaD format (Introduction, Methods, Results, and Discussion) reflects classical arrangement.

However, unlike the pathos found in classical theory, scientific argumentation succeeds through rational commitment based upon agreed-upon procedures. Denying that Popper's theory of "falsification" is the only force at work in scientific theory, Gross (1990) showed the inductive characteristic of science (p. 41). Yet both induction and deduction have their shortcomings, he said, asserting that scientific certainty is a myth (p. 96). Instead, scientific world views are accompanied by "rational conversion" (p. 98), that is, empirical evidence and data supported by rhetorical argumentation.

Gross (1990) also described the importance of authority and power relationships in science. He cited as evidence the competitiveness of science during the emergence of the Royal Society in conflicts among scientists claiming to be the inventors or originators of an instrument or proposed theory. The journal publication date (that is, written discourse) soon became the means of establishing one's priority. Secondly, he used speech act theory to show the power of peer review in preventing true speech acts. Publication (that is, written discourse again) is the means by which review and consensus determine scientific knowledge.

Finally, he pointed out the link between science and society. Citing the controversy over recombinant DNA experiments, he defined two opposing ideologies that control the content of research programs in science in very different ways—technical ideology, which argues for the independence of science and research, and social ideology, which acknowledges the need for society to place limits on science for the protection of society and the environment. In sum, Gross (1990) advocated the relativity of science and the predominance of rhetoric in science, which ultimately uses persuasion to advance its knowledge.

In another *Rhetoric of Science*, Lawrence Prelli (1989) analyzed how scientific communication is created and evaluated rhetorically. In doing so, he used rhetorical invention in science to show that "arguers in science continuously use an identifiable, finite set of value-laden topics as they produce and evaluate claims and counterclaims involving community problems and concerns" (p. 5). Employing the description of scientific paradigm by Thomas Kuhn (1970), Prelli (1989) asserted the rhetorical quality of scientific discourse and its informal logic. He showed that science is persuasive, and thus rhetorical, through its symbolic inducement, whereby models, values, facts, and sense of community are transmitted to adherents of mature science in the following ways: through established paradigms studied and imitated during scientific training; through newly formed paradigms created during revolutionary periods in science; through specific contextual situations involving variables such as financial assistance, adequate research facilities, and interests of colleagues; through its being addressed to audience; through its need to be reasonable, that is, proposing to solve problems; and through its being invented discourse.

This last persuasive feature of scientific discourse, its inventiveness, enables rhetoric to contribute to the creation and progression of scientific discourse. Having established the reasonableness of scientific discourse, Prelli (1989) first showed how the scientist-rhetor must determine reasonable rhetorical goals. The scientific audience perceives as reasonable those researchers who pursue legitimate scientific objectives, "identifying, modifying, or solving problems that bear on a specific scientific community's maintenance and expansion of their comprehension of natural order" (p. 122). Aims that advance scientific order are those which deal with questions of whether or not something exists, of the theoretical meaning or scientific value of claims, or of methodological procedures.

He then proposed the use of scientific *stasis* procedures for determining both how scientists can evaluate adverse community criticism and how they can formulate effective responses to criticism. Presented are four "superior *stases*" (p. 145), which can identify points of argumentation—evidential (questions of the existence of evidence), interpretive (of scientific constructs and information), evaluative (of scientific significance), and methodological. The classical *stases* can then be applied in turn to each of these superior *stases*, giving a set of sixteen questions dealing with conjectural, definitional, qualitative, or translative issues.

Prelli (1989) next used rhetorical invention to show how to address the questions revealed by *stasis* procedures by proposing a set of scientific *topoi*, which guide scientists' lines of argument. After earlier proposing *topoi* that describe scientific ethos (universalism, communality, disinterestedness, and organized skepticism, and their contraries), he expanded his listing to include problem-solution, evaluative, and exemplary *topoi*. Under problem-solution *topoi* are lines of argument dealing with explanatory, predictive, and taxonomic issues; scientific anomaly; experimental and observational competencies; replication and corroboration; and precision of measurements. *Topoi* that provide evaluative arguments are those dealing with accuracy; internal and external consistency; scope of research; simplicity of presentation; ability to provide future research topics; and a somewhat anomalous elegance *topos*, which Prelli asserted at work when scientists judge theories by their neatness and balance. Finally, his exemplary *topoi* of example, analogy, and metaphor come into play when scientists need to conceptualize phenomena.

Marcello Pera (1994) synthesized Aristotle, Perelman, and Kuhn, among others, to propose a dialectical model for scientific argumentation, which includes scientific rhetoric as a central component. The pivotal concept in *The Discourses of Science* is that "the scientist arrives at universal principles by collecting data and opinions, discussing rival views, and convincing his interlocutors that one view is better than another" (p. 50). To derive this merging of facts, method, rhetoric, and dialectic, Pera first showed the limitations inherent in the Cartesian model of science, which posits that the absence of method leaves science non-cognitive and irrational, a state which Pera refers to as the Cartesian syndrome. The paradox of holding such a methodological model is that science itself suffers—or, in Pera's words, is destroyed—from the application of hard-fast methodological rules. Pera showed that the more precise the scientific method, the less adequate it becomes, and that scientists have acknowledged their need for vague rules.

The fact that science requires an imprecise methodology presupposes the need for rhetoric. Specifically, scientific rhetoric assists scientists in choosing, interpreting, and applying a specific methodological rule; justifying initial premises; and supporting or discrediting scientific hypotheses. In defining rhetoric, Pera (1994) adapted Perelman's merger of the Aristotelian concepts of persuasion and dialectics whereby attacking and refuting are part of a scientist's means of persuasion. Consequently, Pera defined scientific dialectics as the logic that validates scientific rhetoric, that is, "the logic of debate" (p. 108), which he divided into both substantive and procedural factors.

To demonstrate his replacement of method with dialectics as the logic of science, he proposed a dialectical model of science, in which science is seen as a game with three players (p. 133)—a group of the scientific community, a rival group, and nature. At the conclusion of the game, a potential change in theory will be decided. Because the game cannot always be won by scientific fact alone, he identified a number of rhetorical strategies used by scientists to determine the viability of rival theories. Among these strategies are means for determining a crucial test and for balancing observation and theory, as well as strategies relying either on the associated merits or on achieved or potential results of a particular theory.

PROFESSIONALIZATION OF SCIENCE

A major goal of this present study of ornithologists' discourse has been to determine the point at which American ornithology was professionalized by analyzing specific features of its discourse. It should then be possible to make generalizations about the written features most characteristic of professionalized scientific discourse in general. In order to determine when a science professionalizes, one must of necessity determine at what point that profession defines itself and its practices as distinctive both from other scientific practices and from non-science. Such expressions of ontology and epistemology are crucial components in the process of the professionalization of any field. Thus, several studies of the professionalizations of specific disciplines and one book-length study of demarcation practices of science are of interest here.

In a study of the professionalization of the field of experimental parapsychology, Paul Allison (1979) described a struggling field attempting to achieve credibility. Though the field created a professional association that supported two journals, set stringent membership and methodological standards, and regularly solicited criticism from established science, experimental parapsychology did not receive the recognition it desired. Major factors inhibiting professional scientific recognition were the inability to replicate phenomena, conflicts with current scientific theories, and, most importantly, the presence of an amateur, or lay, element of parapsychology (that is, the occultists, astrologers, mystics, etc.), which was anathema to empirical science. Exacerbating the situation was the inability to obtain federal grants and publish in prestigious scientific journals. Though half of professional parapsychologists worked in academia, most had little contact with graduate students, a major source of recruitment for professional disciplines.

Thomas Gieryn, George Bevins, and Stephen Zehr (1985) used the two famous creation/evolution trials to investigate science's views of itself as a profession. In the 1925 Scopes trial, science distinguished itself from religion by its ontological, epistemological, methodological, functional, and social differences. In the 1981 Arkansas trial, science argued successfully to exclude pseudo-scientific views from public education by describing science as skeptical, communal, meritocratic, and disinterested—characteristics already identified by Robert Merton. In effect, science employs a market model in its arguments for professionalization: "control the market for knowledge by demarcating their profession and its distinctive commodities from competitive producers" (Gieryn, Bevins, & Zehr, 1985, p. 394).

Henrika Kuklick (1980) described the problems the field of sociology has encountered during its attempts at professionalization during the twentieth century. Though it became academically institutionalized and supported professional journals, it has remained plagued by a number of theoretical and substantive issues.

British astronomy, as personified in the Royal Astronomical Society, effectively professionalized during the last decade of the nineteenth century. John Lankford (1981) described the founding of the British Astronomical Association in 1890 as an organization focused on the needs of the amateur. Soon afterward, the Association's first president, Captain William Noble, recognizing the declining status of the amateur astronomer, announced that the Association's members would "specialize" by organizing themselves into observation sections for studying individual planets and stars. The resulting data would then be shared with professional astronomers. In essence, the creation of the Association relegated amateurs to simple observation and left professionals with scientific activity, thereby institutionalizing amateurs within the discipline. Lankford described a long-standing controversy between amateur and professional astronomers over the value of using small versus large telescopes. With the division into amateur and professional branches, disagreements like this one soon ceased because authority roles were now delineated.

Jean O'Connor and A. J. Meadows (1976) found that the professionalization of British geology was delayed until the mid-twentieth century. An early division into amateur and professional camps occurred in 1858 with the formation of the Geologists' Association as an alternative to the Geological Society of London, which had become too specialized for those only casually interested in geology. By 1900, university training for geologists was established; thus, research and publication were becoming increasingly important. By the early twentieth century, the educational gap between both groups was evident. Professionalization did not begin to occur, however, until mid century. O'Connor and Meadows cited two reasons for the delay. First, since geology was an observation-based science, amateurs could still participate in a major way in geological research. More importantly, the number of professional geologists in Britain was too small to support professional journals until the end of World War II; thus, amateurs were needed to keep these publications afloat financially. Similar developments occurred among the regional geological societies in Britain, though there were more clashes between amateur and professional. In general over the years, amateur geologists in Britain were most dissatisfied with the increasing incomprehensibility of geological journals. O'Connor and Meadows attributed this displeasure to standardization of formats which emphasized reviews of research, the refereeing of manuscripts, and the need to publish original research.

Of necessity, professionalization implies creating boundaries—lines of demarcation—between amateur and professional. Thus, Charles Alan Taylor's *Defining Science: A Rhetoric of Demarcation* (1996) is relevant for this present rhetorical study, which interrelates historical, social, and linguistic components of ornithologists' discourse in order to posit a time frame for professionalization. Taylor argued for the priority of contextual and social influences in distinguishing science from non-science, that is, by placing such distinctions in a rhetorical context. His goal, as he defined it, was to "illustrate how contextual boundaries between science and other practices are constructed" (p. 81).

He first showed that the job of defining science as such has been in the realm of philosophers of science who have traditionally searched for prescriptive answers to what constitutes scientific knowledge. Verification and falsification have been central positions upon which philosophers of science have relied. However, these attempts to demarcate science rely on individual interpretation on the one hand and potential claims of unreliability of results on the other. Other philosophers have used as a defining point the concept of scientific method, or "method talk," as Taylor (1996) put it. But, as he showed, there is not just *a* method, but rather a number of methods, each of whose operation depends upon specific social contexts. Given descriptions both of shared assumptions and values by Kuhn and others and of Toulmin's concept of "ideals of natural order" in guiding science, the contextual component of any definition of "science" becomes apparent. Affirming Steve Fuller's social-epistemological view of science, Taylor argued for a view of science as a set of social practices which are constituted by and for those maintaining the role of "scientist."

Both social and historical studies of science reaffirm science as socially and contextually situated. Sociologists of science have traditionally highlighted those social structures of science that enable the scientific community to function successfully. Robert Merton's description of the ethos of science, that is, its normative structure, may be deemed the key concept derived from sociologists of science. Harriet Zuckerman has refined Merton's description by identifying two distinct norms: (a) cognitive, including methodological procedures; and (b) moral, involving scientists' attitudes and behavior. Practically speaking, rhetoric enters into the picture whenever challenges to these norms arise. An extreme relativist view interprets even scientists' standards as rhetorically negotiated. Within sociology of science, Taylor (1996) identified two main branches. The strong program attempts to place science in an objective framework in order to facilitate the sociological study of its collective practices. In contrast, the constructivist program focuses this research strategy in specific laboratory settings, with special emphasis on the evolving discourse resulting from ongoing laboratory investigations. In fact, for rhetoricians a key derivative of sociological studies of science is the value of studying scientists' discourse, both written and spoken. Historical studies of science, on the other hand, provide evidence for or against philosophical accounts of scientific epistemology. In particular, historical narratives enable rhetoricians to identify communal relationships among scientists and are useful in studying interdisciplinary controversies, as well as science in the process of professionalizing (pp. 95–96).

Having integrated rhetoric into the major components of the literature of science, Taylor (1996) suggested that demarcating science rhetorically by studying its discourse is an effective way of distinguishing not only science from non-science, but one specialization from another. In particular, a view of science as an ecosystem, that is, an entity contextually situated not only in private but also in public forums, will facilitate such an analysis. Because rhetoric is addressed discourse, the nature of the scientific audience is critical. Arguing that too often a communal, archetypal scientific audience is assumed, Taylor turned to a description of the contextual nature of science for a more accurate key to audience. Criticizing as too narrow those internalist perspectives of science arguing from within prevailing scientific contexts, Taylor favored a view of science that is influenced by broader social and cultural practices. He questioned the focus of previous studies of science on the nature of science and scientific facts and on the various spheres of argument from personal to public to technical. Instead, demarcating science by studying the discursive practices and rhetorical processes of science is a more fruitful endeavor for rhetoricians of science.

Finally, he implemented his rhetorical strategies in a study of two recent controversies: the inclusion of creationism in public school science curriculum and the cold fusion controversy. In his view, demarcation strategies of traditional science have been largely responsible for the inroads of creationists into public education. By demarcating science as an open-ended, uncertain endeavor based on community-accepted norms such as probability, falsification, and disinterestedness, traditional science set itself up as an exclusive, adjudicating body in matters of science. However, creationists were able not only to proclaim themselves as degreed scientists, but also to argue based on empirical observation, strict induction, and common-sense realism. Because popularizations of science have given the public a view of science as certainty and because traditional science insulated itself from the public realm, creationism has continued to maintain its relevance to the general public, thus threatening the ultimate authority of traditional science.

In the cold fusion controversy, Taylor (1996) used Merton's norms of disinterestedness, communality, organized skepticism, and universalism to trace the demarcation of science. In the process, he showed how these norms were contextually situated, and thus rhetorical. For instance, two scientists working in a basement laboratory at relatively modest expense were a catalyst for heightened negative responses from recognized fusion researchers, who had already spent 45 years and multimillions of dollars in ongoing research. Second, physicists, who claimed disciplinary rights to fusion research, foresaw an interdisciplinary turf battle for economic and political power, as well as disciplinary integrity, with the two chemists claiming cold fusion results. Finally, the need to make significant changes to current government funding and peer review structures contributed to a negative response to further cold fusion research by the federal government. Most importantly, by using a diverse cross-section of the discourse generated by the creationist and cold fusion controversies, including news media accounts and associated editorials; transcripts of government hearings; histories of the controversies; technical reports; and articles and commentaries in professional journals, Taylor showed science as social practice carried on in a discursive, and thus rhetorical, environment.

LINGUISTIC STUDIES

In analyzing specific features of ornithologists' writing, it is crucial, of course, to understand the linguistic choices these writers made. As has been shown in numerous studies, and will be reiterated in this study, a key feature of scientific discourse is its technicality, identified most explicitly by the technical vocabulary science creates. Scientific discourse must also have sufficient resources to argue effectively. Nowhere is this more important than in its ability to support its claims, and to do so scientists must be able to choose from a wide range of vocabulary in order to present, and indicate the importance of, the evidence it has gathered in support of its claims. Two recent linguistic studies provide insight into analyzing these linguistic features of ornithologists' writing.

In *Writing Science*, M. A. K. Halliday and J. R. Martin (1993) and their co-contributors used systemic functional linguistics to highlight the lexico-grammatical features of scientific writing. This methodology enabled the authors to study scientific writing from two perspectives: language and social context. The overriding theme of the book is that the evolution of scientific grammar made possible the evolution of scientific thought. One reason for this grammatical evolution was the ability of language to nominalize and create polynomials. These forms became a powerful means to condense information, convert events into objects, and minimize negotiation of ideas, thereby giving scientific writing a privileged position. Secondly, the birth of grammatical metaphor, whereby multi-nomials enable one phenomenon to represent another, facilitated the embedding of scientific processes. Additional linguistic features that are shown to account for the uniqueness of scientific discourse are its interlocking definitions, technical taxonomies, lexical density, syntactic ambiguity, and semantic discontinuity.

Linguistic choices also indicate levels of expertise, or "attitudes toward knowledge," as described by Wallace Chafe (1985, p. 271), who termed as "evidentiality" several categories of linguistic expressions reflecting speaker/writer attitudes. Chafe described the following eight evidentials:

- degrees of reliability—signaled by adverbs like *maybe*, *undoubtedly*, and *possibly*
- belief—with expressions such as *I think* and *I suppose*
- induction—by *must* and *evidently*
- sensory evidence—by expressions like *I see*
- hearsay evidence—as for example *it seems* and *apparently*

- deduction—signaled by *presumably* and *could*
- hedges—like *sort of* and *about*
- expectations—including expressions like *oddly enough* and *of course.*

An analysis of the use of such expressions shows interlocutors' own attitudes toward the information they are giving and signals the amount of authority listeners can attribute to the information they are receiving. In effect, the linguistic expressions of evidentiality used by experts will vary considerably from those chosen by the amateur. Thus, they are effective means to delineate each type of discourse.

DISCOURSE STUDIES

Since this study of American ornithologists' writing analyzes discourse from the merger of amateur and technical ornithologist in 1883 through the field's professionalization during the twentieth century, studies comparing the writing of amateurs, professionals, and popularizers also reveal the possibilities that a study of ornithological discourse can achieve. A number of researchers have described characteristic features of both expert and novice writers, and one book-length study has described the specific choices made by five biologists as they revised for both professional and popular audiences.

Michael Carter (1990) turned the general knowledge of the novice and the local, domain-specific knowledge of the expert into a continuum in which general strategies of writing vary in generality, becoming more context-bound as writers through experience and contact enter more fully into their discourse community which houses domain-specific knowledge (p. 279). Amateurs/novices need general heuristic strategies to enable them to imitate expert discourse, whereas expert writers through their experience in contact with their specific community do not ordinarily use general strategies for writing but depend on know-how and intuition (pp. 271–272). However, Perkins and Salomon (1989) have shown that experts, when confronted with unusual problems in their disciplines, turn to more generalized, though clearly contextualized, strategies (p. 20).

Analyzing novice writers' attempts to enter academic discourse, David Bartholomae (1985) acknowledged expert writers' recognition of their privilege to speak. Unlike amateurs, experts locate themselves in a discourse community with projects and agendas that limit the discourse itself. Experts, consequently, write from an imagined position of privilege or authority (p. 139), a privilege not only of being inside the discourse community, but the privilege of "being granted a special right to speak" (p. 143), a special right that evolves from the ability to place one's discourse in opposition to some "common discourse" (p. 156) against which it argues. In like manner, Patricia Bizzell (1982) evaluated the writing of experts as successful because they are writing for a discourse community whose rules, major concepts, research goals, and worldview are familiar to them and "in which they know their work can matter" (p. 232).

The protocol analysis of Linda Flower and John R. Hayes shows the influence of experience and contextual awareness in the success of expert writers. For instance, Flower (1989) distinguished expert from amateur by experts' ability to integrate their own writing goals (personal, cognitive influence) with contextual goals (social influences via a discourse community) (p. 292). The study by Flower and Hayes (1980) of subjects writing for a *Seventeen* magazine audience found expert writers more aware of the entire rhetorical situation, which experts analyzed based on their own experience, a skill evidently lacking in amateur writers. Experts also tapped into formats and organizational structures previously stored in memory and were able to develop topics in more depth. Experts were better idea-generators because they were more aware of affecting a reader (pp. 26–30). These findings reflect again the expert writer's knowledge of a particular discourse community: both the individual members themselves and the community's established conventions.

Like Charles Bazerman (1988), Greg Myers (1990) studied the social forces at work in scientific discourse, but restricted his analysis to the writing of five American and British biologists. *Writing Biology* uses specific reports, proposals, articles, popular writings, and one book to show that scientific texts do not just reflect, but help create, scientific knowledge. In particular, Myers (1990) compared original and revised versions of texts, texts and reviewers' comments, and texts written for varied audiences.

His first case study was that of two evolutionary biologists writing proposals. Shown here is the consensus-building needed to produce a successful proposal. Though the two biologists made few substantive changes to their research plans, they did make many revisions to enhance their personae as researchers, to relate their work to past research, and to show the relevance of their work to the contemporary research program. The paradox here is the need to show that their research is new, yet relevant to the ongoing research program. Thus, both researchers had to change the context of their projects in order to accommodate consensus. Myers (1990) pointed out also that approval of such research proposals enlarges—in effect, redefines—the research program itself.

His second case study, that of two biologists' research articles, emphasized the changing status given to scientific claims during the review and revision process. These negotiations affected the form of the text, rather than the claim itself or its data. That is, the biologists had to show that their articles were relevant to the journal and situated in the literature as a whole. In the process of publication, both had to lower the status of their claims. Among major changes to format and style were revisions to accommodate a particular journal format and changes in diction to minimize persona and project a less assertive tone, thereby making claims more tentative. In other words, like the proposal process, the publication process succeeds through consensus building.

Myers (1990) next used a sequence of five articles to show the importance of narrative in a biological controversy. He posited three levels of narrative through which scientific argumentation evolves: from a narrative of the research field, to that of the research project, and then of the specific empirical study. In the process of constructing, interpreting, and reinterpreting these narratives, the argument becomes so complex and interwoven that a successful refutation of one aspect of the argument can invalidate the entire argument. He stressed the significance of narrative in that "biologists do bring in evidence, but it is effective, or isn't, because of the way they make the whole story fit together so that it has meaning" (p. 103).

Finally, he focused on popularizations of science. He interconnected narrative and popularization by showing how four professional biologists in the process of rewriting their research for semi-professional magazines had to restructure their narratives of science into natural history narratives. He used external features of discourse, such as titles and illustrations; conceptual reorganizations in abstracts and introductions, as well as in the body of the texts; and decreases in sentence complexity, passive voice, and technical terminology to show how the research project and scientific methodology give way to simple observations of objects in nature. The study concludes by using E. O. Wilson's *Sociobiology* in order to show the same process in the opposite direction: that is, Wilson's restructuring of natural history narratives into narratives of science.

With these studies in mind, let us now turn to an analysis of American ornithological discourse. Chapter 2 provides an historical overview of the 300 years preceding the birth of the American Ornithologists' Union and its journal, *The Auk*. Evident from this overview are two lines of development: natural history and experimental science. Chapter 3 then offers discourse models representative of these two lines of development. Chapters 4 and 5 focus on *The Auk* by tracing the progress of natural history and experimental science discourse over the past century of American ornithology, by suggesting a third discourse model representative of an evolving discourse, and by offering a set of indices representative of professionalized discourse. The study concludes by placing American ornithological discourse within the rhetoric of science.

2

The Emergence of American Ornithological Discourse: 1600 to 1900

This study focuses on the development of American ornithological discourse from natural history to experimental science using American ornithology's professional journal, *The Auk*, which first appeared in 1884. But ornithological discourse in America did not arise in a vacuum, nor appear out of nowhere. What happened to ornithological discourse before the 1880s? What traditions led ornithologists to write the way they did and discuss the topics they did? This chapter answers these questions by showing the development of ornithological discourse from Renaissance craft literature in England to natural history study in America to an initial realization of American ornithologists' professional self-image[1]. This progression is not a neat continuum but a parallel, interwoven development of both popular interest in birds and a deeper scientific probing into their origin and evolution. As shown in Figure 2.1, this chapter traces the discourse arising out of these two lines of development: on the one hand, the popular discourse from craft literature to bird listing to natural history narratives; and on the other hand, the technical discourse from avian taxonomies to discourse such as manuals, scientific texts, and handbooks, which indicate a developing sense of professionalism. These two discourses are not separate but interrelated, with ornithologists such as

Audubon influencing both lines. Such interplay of personages and discourse in the early development of both natural history and technical ornithology provided the conditions for the continued intermingling of amateur and professional, which continues to influence the discourse of American ornithology, as this study demonstrates. Thus, an understanding of American ornithological discourse rests in an awareness not only of the social interactions and scientific developments of the late nineteenth and twentieth centuries, but also of the topics and forces guiding the evolving discourse during its earlier development.

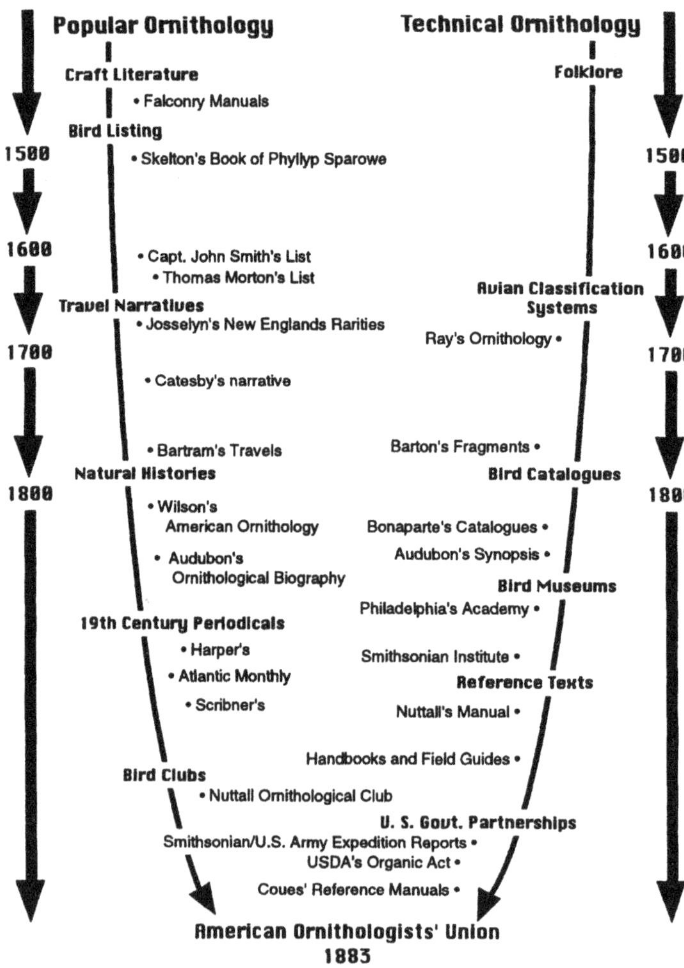

FIGURE 2.1. American ornithology's historical roots.

Craft literature focused on the utilitarian and entertainment value of birds, popularizing them so that they became of interest not only for their culinary and recreational use, but also for their own sakes as objects of study and observation. Interest in bird listing in Renaissance England and colonial America showed that people had begun to study birds as part of their daily lives; thus, the love affair with birds had begun. The bird accounts found in eighteenth-century travel literature, which had expanded writing about birds from simple area lists to natural histories, reflect this increasing interest in birds. Yet these writings fail to account fully for what was to become professional ornithological discourse because they were imbued with an emotionalism and awe in their attempt to describe the bird life in America. Influencing this viewpoint also was the mere lack of awareness of "birds" as a science, quite unlike the recognition given to the study of plants or human anatomy (see Tebeaux, 1997, pp. 14–19). Through the mid-nineteenth century, American ornithology was in its "discovery phase," a period devoted to the description and classification of birds. Artists Wilson and Audubon were key figures in the search, adding a pictorial realism that previous writers could not provide, though these authors also reflected the anthropomorphic views of their predecessors. Handbooks and guides at this time were often inaccurate and poorly illustrated. Reflecting on American ornithology before the 1850s, ornithologist Charles Foster Batchelder (1937/1974) commented that "ornithology was still a simple subject,—hardly a science,—species as yet were fixed entities, unchanged and unchangeable from their beginning, and the possibility of geographical variation within a species not dreamt of" (p. 9). In order to correct these shortcomings, classification and taxonomy became central to the emerging ornithology of the mid-nineteenth century, culminating in the extensive scientific research conducted by Spencer F. Baird and the Army medics, whose research produced a major report on Western birds and numerous Western field guides. Only after this phase was completed and Darwin's impact was felt could ornithology turn to geographic variations in species and the climatic laws governing this variation, a popular topic among ornithologists like Baird and his student Robert Ridgway during the remainder of the nineteenth century. Only at the end of the century with the appearance of *The Auk* did American ornithology begin to organize itself into the specialized field we know today. Yet, the priorities of early American ornithologists were not the behavior of birds and their relationship to their habitat—topics that appeared frequently in amateur ornithologists' works, as for instance in the writings of transcendentalist Henry David Thoreau and naturalist John Burroughs. Rather, such topics were delayed in the technical literature until the early twentieth century. This interplay between amateur and professional and their eventual rift will be the focus of this study of American ornithological discourse. For its part, this chapter traces the development of the major forces of natural history and taxonomy, the participation of the amateur in the study of birds, and the formative period when ornithology and its discourse were beginning to provide the milieu for the creation of what would become the American Ornithologists' Union.

CRAFT LITERATURE

De arte venandi cum avibus (Wood & Fyfe, 1943), a medieval falconry manual written during the first half of the thirteenth century by Roman emperor Frederick II, is often cited as the birth of ornithology, in particular because the manual also contained both physical and natural history descriptions of birds. Similarly, in Renaissance England, instruction manuals devoted to the art of falconry were among the earliest forerunners of ornithology. Though these manuals may strictly be called technical writing rather than scientific writing, "science" changed radically during the Renaissance, thereby blurring the distinction between the two types of discourse (Tebeaux & Killingsworth, 1992, p. 7). As Elizabeth Tebeaux (1997) has pointed out, "[English Renaissance] books, written to help individuals correctly perform processes, anticipate modern scientific works with their emphasis on method and procedure" (p. 3). Thus, Renaissance falconry manuals inform our understanding of the emergence of ornithological science. Like later American ornithological discourse, these manuals contained descriptions of plumages, taxonomic categorizations, and behavioral features of birds—topics that have become important to modern ornithology. More importantly, the art of falconry popularized this gentleman's sport in England, in much the same way that natural history writing, which emerged in the eighteenth and nineteenth centuries, has popularized ornithology in more modern times.

By the early seventeenth century, falconry's popularity in England had transformed the pastime, also called hawking, into a national sport, in which hawks symbolized the falconer's position in the country's hierarchical society. Each rung of Renaissance society had its hawk, from the king's gyrfalcon to the yeoman's goshawk to the clergyman's sparrow hawk. Even the servant was recognized by the gentlemanly event, which permitted him to fly the diminutive kestrel, useless though it was to falconers since it hunted only small land birds, lizards, and frogs. Thus, falconry manuals had their own unique rhetorical and linguistic purpose in English society, a role evident in the many references to hawking in Renaissance literature (Hieatt, 1983). Falconry manuals taught readers the language of the sport, initiating them into the terminology—a gentleman's vocabulary—valuable for admittance into the gentry, though in the peasant's case only symbolically. Thus, by the end of the Renaissance, knowledge of falconry had become significant enough to warrant authors appending falconry terminology to their works.

The structure of Renaissance falconry manuals was generally the same: First were definitions of hawks followed by explanations of capturing, taming, and training for hunt; then general instructions in the care of hawks; and finally diagnosis and cure of hawks' diseases. The earliest such English treatise appeared in 1486. Representative texts which spanned the period of falconry's greatest popularity in print were those by George Turberville (1575), two early seventeenth-century works by two falconers Symon Latham (1615, 1618) and Edmund Bert (1619), and a condensed section on falconry printed by Richard Blome (1686). The multiple editions of these and other Renaissance falconry manuals indicate their popularity.

In his analysis of the development of scientific knowledge, Thomas Kuhn (1970) argued: "Scientific development becomes the piecemeal process by which [facts, theories, and methods] have been added, singly and in combination, to the ever growing stockpile that constitutes scientific technique and knowledge" (pp. 1–2). The connection between craft literature, such as falconry manuals, and scientific discourse can be seen from this perspective; for these manuals likewise were merely piecemeal compilations from previous authors' works. As Kuhn concluded, such

> early fact-gathering is a far more nearly random activity than the one that subsequent scientific development makes familiar. Furthermore, in the absence of a reason for seeking some particular form of more recondite information, early fact-gathering is usually restricted to the wealth of data that lie ready to hand. The resulting pool of facts contains those accessible to casual observation and experiment together with some of the more esoteric data retrievable from established crafts like medicine, calendar making, and metallurgy. (p. 15)

In this manner, falconry writers collected tidbits of advice on the care and training of hawks and the curing of their diseases. Some of these writers, such as Bert and Latham who were themselves falconers, often added their own methods and home remedies to the collection. These works then are allied more with the natural history end of ornithological discourse, due to the more subjective, personal nature of the manuals.

Missing from these texts were analyses, methodology, and experimentation—that is, the scientific perspective necessary to consider them the beginnings of ornithological discourse in English. Their significance here is that they included many of the same topics of future importance to ornithology. Elements of systematics appeared as, for instance, in this excerpt by George Turberville (1575), who first divided hawks into nine general kinds and then subdivided them functionally by the manner in which they kill their prey:

> The ſeauenth is the Tunician Falcon. These feue kinds of Falcons, are al hardie and good, according to the pray that belongeth to their force and nature. For they are diuers in nature, and of ſeuerall plumes, and beſides that, dispersed and eyried in divers countries. Wherefore, I meane to touche euery one of their natures, and to declare you, whereunto eche is enclined, according as they are maned, and gouerned, and firſt of all I will treate of the Falcon Gentle.
>
> But before I ſpeake of the Falcon Gentle, I will ſomewhat touch the Etimologie of the name, and lay downe the opinions of two learned men, and auncient writers as touching that matter. (p. 27)

Seen also were the beginnings of taxonomic disagreements among authors, which has persisted throughout ornithology's development. For instance, Latham criticized Turberville's divisions, which placed haggard falcons first in hawk hierarchy. Blome first divided hawks into long-winged and short-winged species before naming specific kinds of hawks, foreshadowing later emphasis on morphology in ornithological description, as did falconry authors' definitions of hawks according to plumage. Finally, two more foreshadowings of future ornithological research were present. Sections on diseases and cures showed interest in the physiology of birds, and discussions of training and care of hawks—including housing, care during molt, attraction to the lure, and methods of bathing and feeding—revealed knowledge of bird behavior.

Also among early craft literature were many books demonstrating the culinary value of birds, as for instance how-to books on cooking and raising game birds, as well as gentlemen's books on hunting and falconry during the Renaissance. British interest in bird accounts spread across the Atlantic Ocean as well. The early colonists of the New World wrote briefly about birds, but the substance of their writings was simple, meager lists of the most common birds seen, with an emphasis on birds used for cooking, often including comments on the taste of various game birds. Typical were accounts by Captain John Smith (1612) and Thomas Morton (1637). These works helped instill interest in birds, not only as utilitarian objects but as subjects for study as well.

FROM READING TO WRITING: BIRD LISTING

Through the centuries, a popular pastime of writers on ornithology has been the compilation of a bird list. These lists have been a common form of discourse through the early twentieth century, as well as a discourse responsible for the continued intermingling of amateur and professional even in recent times. From medieval writer Geoffrey Chaucer (1375–1385/1980) came *Parlement of Foulys*, describing the major characteristics of 37 species.

Soon afterward, John Skelton (1508/1983) incorporated what was essentially an early bird list of 69 species into his poem *The Boke of Phyllyp Sparowe*. In 1544 William Turner published his own Latin interpretation of the birds referenced by Aristotle and Pliny, a work which was still referenced 100 years later in John Ray's important English work on birds (E.G. Allen, 1951, pp. 415–416; O'Malley, 1968, p. 26). There were also a number of early English county histories, as for example Richard Carew's *Survey of Cornwall* (1602), which included a discussion of game birds, hawks, and singing birds of southern England, and the bird list by Sir Thomas Browne (1684) (E.G. Allen, 1951, p. 417). Edward Topsell's fragmentary *Fowles of Heauen* (c. 1614/1972) was a compilation from various sources, as Kuhn (1970) describes, which included accounts of American as well as European birds accompanied by 124 colored woodcuts. His descriptions of 17 American birds, including at least one obtained through Richard Hakluyt, appear to be among the earliest accounts of American birds following the establishment of Jamestown (Christy, 1933). Later travel works also contained bird lists, for instance, those by Captain Jonathan Carver (1778), William Bartram (1790), and Gilbert Imlay (1793). Also clarifying the American bird record were state histories such as Jeremy Belknap's on New Hampshire (1792), Thomas Jefferson's on Virginia (1788), and St. John de Crevecoeur's on the bird life of New York (1782). Though of seeming insignificance, late eighteenth-century bird lists such as these have been described by ornithological historian Elsa Allen (1951) as "'germs' of modern descriptive ornithology . . . the nucleus around which better and more definitive writing has clustered" (p. 524).

NATURAL HISTORY FOR ITS OWN SAKE

Craft literature, including falconry training manuals and culinary books, as well as the making of bird lists, aroused interest in birds during the sixteenth and seventeenth centuries. In the seventeenth and eighteenth centuries, interest in the natural history of birds began to appear as travel writers described the New World they were experiencing first-hand. Ultimately, written descriptions of casual observations were not enough to satisfy the craving for knowledge about America's bird life. Hundreds of color illustrations by Wilson and Audubon captured the attention of nature enthusiasts, as writing about birds spread into the popular literature.

Natural History as Travel Narrative

With the exploration and colonization of the New World came the advent of New World travel narratives, motivated by the fascination in Europe for everything associated with the New World. Learning about the New World satisfied Europeans' spirit of discovery and their desire for the new and mysterious. The rise of empiricism motivated this desire to concretize by putting experience into words and pictures, although, as was typical of these narratives, emotionalism and anthropomorphisms abounded because writers were unable to distinguish objectivity from aesthetic appreciation. Part of this new experience was American flora and fauna; thus, among the topics of colonial travel writers were descriptions of American birds. Later travel writers, such as Mark Catesby and William Bartram, became significant figures in the development of American ornithological discourse; their colonial accounts, along with the work of artists Wilson and Audubon, became the cornerstone from which nineteenth-century American ornithological science emerged. However, seventeenth-century travel narratives were not as influential. John Josselyn's *New Englands Rarities Discovered* (1672), the first natural history account of the New World, included only eleven bird entries, covering a mere six pages, as for example the following entry, a narrative of the hummingbird:

> The *Humming Bird*, the leaſt of all Birds, little bigger than a *Dor*, of variable glittering Colours, they feed upon Honey, which they ſuck out of Bloſſoms and Flowers with their long Needle-like Bills; they ſleep all Winter, and are not to be ſeen till the Spring, at which time they breed in little Neſts made up like a bottom of ſoft Silk-like matter, their Eggs no bigger than a white Peaſe, they hatch three or four at a time, and are proper to this Country. (pp. 6–7)

Since Josselyn emphasized the practical uses both of birds' feathers and of the birds themselves for food and medicinal purposes, bird descriptions were minimal.

From the perspective of scientific ornithology, the most important travel narratives of the eighteenth century were those by botanists Mark Catesby and William Bartram. Catesby's *Natural History* (1731–1743), the first truly accurate natural history account of America, contained 100 color plates and over 100 pages of text about birds, as well as plants and other animals. Catesby's paintings of 109 species of birds were, for the most part, excellent representations. More significant for ornithology was botanist Catesby's interest in the flora of America; for he knowledgeably associated the birds he drew with their environment, depicting them either among typical vegetation or food sources. This association between bird and habitat was a major contribution to ornithology. Also significant from the point of view of ornithological discourse were his detailed textual descriptions, here of the plumage features of the now extinct Carolina parakeet:

> This Bird [The Parrot of Carolina] is of the bigneſs, or rather leſs than a Black-bird, weighing three ounces and an half: the fore-part of the Head

> Orange-colour; the hind-part of the Head and Neck yellow. All the reſt of the Bird appears green; but upon nearer ſcrutiny the interior vanes of the moſt of the wing-feathers are dark-brown: the upper-parts of the exterior vanes of the larger Wing-or-Quill-feathers, are yellow, proceeding gradually deeper colour'd to the end, from yellow to green, and from green to blew: the edge of the Shoulder of the Wing, for about three inches down, is bright Orange-colour. (p. 48)

Due to his scientific inclination, he included empirical details of habitat, behavior, measurements and weights, and descriptions of song and color. Though birds' names were not standardized, the names Catesby ascribed to his plates were accurate enough that Linnaeus used 71 of these plates in codifying his binomial nomenclature (E.G. Allen, 1951, p. 465). As a result, the names of 59 species are attributed to Catesby (Stevens, 1936, pp. 206–207). As America's first ornithologist, he "laid the foundation of popular ornithology inaugurating the combination of attractive lifelike portraits with appropriate environment, and short accurate accounts—the beginning, on a small scale, of Wilson's and Audubon's system of illustrated ornithological life histories" (E.G. Allen, 1951, p. 474). Perhaps more significant for this study, his work intermingled natural history study and technical ornithology two centuries ago. As will be seen, this interplay represents a persistent association throughout American ornithology.

William Bartram's main contribution to ornithological discourse was the catalogue of North American birds he included in his *Travels* (1791). Though his descriptions of 215 species contained some mythology as well as ornithological inaccuracies, particularly in the confusion of immature birds for separate species, his account was invaluable for its descriptions of behavior and migration, in addition to its comprehensive bird list. Bartram was "the first scientific American ornithologist" because of his careful, long-term record keeping, his refutation of previously unsupported theories, and his scientific bent, as for instance his dissection of migratory birds in order to study their sexual development (E.G. Allen, 1951, pp. 539–541). Bartram left his mark on ornithological history both as the first to describe several North American species and as a pioneer in the study of bird migration in America. He was the first to show the importance of studying the living bird, even the most common one, not by using a gun, but by observing the bird's behavior. His novel practice foreshadows questions over the priority of the living bird versus the study specimen—questions that will ultimately help to separate the amateur and professional into conflicting camps.

Fueling another conflict between amateur and professional, between natural history and technical science, was Bartram's anthropomorphic and emotional style of writing, as the following passage demonstrates:

> The sun now below the western horizon, the moon majestically rising in the east; again the tuneful birds become inspired; how melodious is the social mock-bird! the groves resound the unceasing cries of the whip-poor-will. . . . O thou Creator supreme, almighty! how infinite and incomprehensible thy works! most perfect, and every way astonishing! (1791/1958, pp. 32, 38)

This style of writing created a controversy during the early years of American professional ornithology, when much of Bartram's work was disavowed and his type descriptions and nomenclature were removed, in a ploy to disassociate emerging professional ornithology from the emotional amateur. Still, nineteenth-century ornithologist-historian Elliott Coues (1903) considered Bartram's *Travels* "the starting-point of a distinctively American school of ornithology" (p. xvii). Also significant was the great influence Bartram had on a prominent nineteenth-century American ornithologist, Alexander Wilson.

Blossoming of Natural History Discourse

Amateur ornithological discourse is much indebted to Scotsman Alexander Wilson, whose immigration to America in 1794 and his subsequent visit to Philadelphia's Museum of Natural Sciences inspired his project to paint all of the birds of North America, as did his contacts with William Bartram, who fostered his ornithological interests. Though Wilson died before he accomplished his goals, he published seven volumes of his *American Ornithology* (1808–1814), the remaining two volumes completed by George Ord[2]. Since Wilson did not conform to taxonomic relationships in the organization of his *Ornithology*, his was not as much a scientific work as one incorporating personal narratives showing both his skill as a field ornithologist and his close contact with nature. Unlike his painter-predecessors, he wrote about the ecology of the birds he illustrated, describing behavior, habitat, and song:

> . . . I have frequently known them [Gold-winged Woodpeckers] fix on the trunk of an old apple tree, at not more than six feet from the root. The sagacity of this bird in discovering, under a sound bark, a hollow limb or trunk of a tree, and its perseverance in perforating it for the purpose of incubation, are truly surprising; the male and female alternately relieving and encouraging each other, by mutual caresses, renewing their labours for several days, till the object is attained, and the place rendered sufficiently capacious, convenient, and secure. At this employment they are so extremely intent, that they may be heard till a very late hour in the evening, thumping like carpenters. (vol. 1, pp. 44–45)

Though he did anthropomorphize, he attacked traditional myths and notions surrounding birds, like swallows hibernating under water (Welker, 1955, pp. 38–39). As Elsa Allen (1951) described him, he was "preeminent at his time, exceeding all his predecessors, and even Audubon, in his ability to describe and interpret the living bird" (p. 553). In the introduction to volume one, Wilson clearly stated his purpose in publishing the volumes: "To furnish instruction blended with amusement, to correct numerous errors by European writers, and to draw the attention of fellow citizens from the discordant jarrings of politics to a contemplation of the grandeur, harmony and wonder of Nature" (as cited in E.G. Allen, 1951, p. 558). He further contrasted his work with many previous ones in that his knowledge came from "personal intimacy" with birds in their natural environment (as cited in E.G. Allen, p. 558). These introductory comments by Wilson foreshadowed the eventual split of American ornithological discourse into popular and scientific camps.

Wilson's contemporary, John James Audubon, also contributed to the natural history of birds. His *Ornithological Biography* (1831–1849), a five-volume accompaniment to his paintings, discussed the behavior and life history of the birds Audubon studied, as in this example:

> Its nest [the blue-winged yellow warbler's], which is singularly constructed, and of an elongated inversely conical form, is attached to several stalks or blades of tall grass by its upper edge. The materials of which it is formed are placed obliquely from its mouth to the bottom. The latter part is composed of dried leaves, and is finished within with fine grass and lichens. The female lays from four to six eggs, of a pure white colour, with a few pale red spots at the larger end. . . .
>
> It leaves Louisiana in the beginning of October. I have never seen the species farther eastward than the State of Jersey, where I killed several within a few miles of Philadelphia, not however until my last visit to that State in 1829. (p. 102)

This perspective was ignored by most of Audubon's European contemporaries who concerned themselves primarily with systematics and nomenclature, topics that only later in the century triggered a sense of professionalism. In all, Audubon described 506 species, most of these with color plates. Robert Henry Welker (1955) compared Audubon's *Biography* to William Bartram's account in that both expressed "an enthusiasm for nature which approached religious awe." Such emotionalism was expressed through diction choices that still presented birds anthropomorphically. Interspersed among Audubon's bird descriptions were sixty accounts which included comments on the wilds of America, social commentary, and character sketches (pp. 65–66, 105).

Popularization Through Natural History

Works such as Wilson's and Audubon's brought attention to American birds like no other works before them. Now, instead of long-winded descriptions, readers could see birds in lifelike form and color. For instance, Wilson's *Ornithology* included 320 figures of 262 species, with color illustrations of most. Audubon's elephant-folio *Birds of America* (1827–1838) contained 435 plates of 497 birds painted in their natural habitat[3]. This ornithological writing and illustration brought the subject of birds increasingly to the attention of the general public, an interest that popular magazines soon took up. Of all the nineteenth-century monthly periodicals, bird articles most frequently appeared in the *Atlantic Monthly*, *Harper's*, and *Scribner's Monthly*, with occasional articles in other popular magazines (Welker, 1955, p. 179). As Welker observed, there were also children's magazines that emphasized bird conservation. The religious, didactic *Youth's Companion* included through 1856 a section entitled "Natural History," which incorporated birds among its animal subjects. A later children's magazine, *St. Nicholas*, also focused on bird conservation articles. Other books on birds published during the century treated birds piously, anthropomorphizing them and continuing to disseminate the lore that ornithologists of the period were trying to correct (pp. 179–187). Also during this time, natural history writers like John Burroughs were publishing accounts of birds' lives in popular periodicals such as *Century* and the *Atlantic Monthly*. Thus, by the end of the nineteenth century, popular interest in birds had blossomed into an ever-increasing following; and natural history as a form of ornithological discourse was now established.

HINTS OF EXPERIMENTAL SCIENCE: THE EMERGENCE OF SYSTEMATICS

Not only had natural history discourse about birds developed during the eighteenth and nineteenth centuries, but the beginnings of a scientific approach to bird study also began to form—an approach that would later become the cornerstone of American ornithology. The focus of this study was the identification of birds both in the field and in the hand in an attempt to understand the relationships among species. It was an interest that would evolve into the discourse of systematics, the attempt by ornithologists to create taxonomic classifications of the family trees of birds in order to trace the evolutionary relationships among species[4]. The study of systematics moved to the forefront of nineteenth-century American ornithology as a direct result of Darwin's theory of evolution, which American ornithologists quickly accepted. Their foresight, in contrast to British ornithologists' half century of hesitation, gave American ornithologists a basis for a wide array of topics for research, which enabled them to formulate the first viable trinomial nomenclature accurately describing avian relationships.

Looking back in ornithological history, evidence of early interest in systematics appeared in an English treatise of the seventeenth century, John

Ray's 1678 translation of Francis Willughby's *Ornithology*. The most comprehensive seventeenth-century handbook on birds, it contained the period's best avian classification system. Willughby, one of the charter members of the Royal Society, compiled *Ornithology* based on his travels with John Ray through England and Europe. The work rejected the folklore of older ornithologies and is considered "the beginning of scientific ornithology in Europe" (E.G. Allen, 1951, p. 424), since it employed structural features such as bill and claw shape, rather than classical Aristotelian principles interrelating form and function, as for example relating bill shape to diet, giving a classification such as seed-eater. It included 78 plates of birds and contained a general account, followed by chapters on land birds and waterfowl, as in the following excerpt, which emphasizes physical, as well as plumage, features:

> It [the North-Country Dunlin] is about the bigneſs of the *Jack-Snipe or Judcock*, hath a ſtreight, channell'd, black Bill, a little broader at the end; oblong Noſthrils; a blackiſh Tongue. The Throat and Breaſt white, ſpotted with black. The middle of the Belly is blackiſh, waved with white lines. The lower Belly, and feathers under the Tail white. All the upper ſide is red, every where ſpotted with pretty great black ſpots with a little white. (Willughby, 1678, Vol. 3, p. 305)

The work retained its position as the most advanced ornithology text for almost a century (Farber, 1982, p. 6) and was the last attempt to offer a complete monograph on birds, because the data had become too overwhelming a task to collate (Stresemann, 1951/1975, p. 45), a circumstance that likewise has affected ornithological writing in the twentieth century. Ray was also one of the first writers to use the term "ornithology" in print. As far as is known, his usage was second only to Fuller, who in 1655 entitled his book *Ornitho-Logie*. The word was ultimately derived from Aldrovandus' Latin usage *ornithologia* in 1599.

Early interest in the classification of birds in America is seen in Benjamin Smith Barton's *Fragments of the Natural History of Pennsylvania* (1799). Through his attempts to classify migratory birds, he coined terms such as "resident species," "accidental," and "occasional visitants" (E.G. Allen, 1951, p. 535), which became part of the vocabulary of technical ornithology, as in this example:

> There is a propriety in the diviſion of the Reſident Birds, into ſuch as commonly continue in the country the whole year through, and ſuch as only occaſionally continue. The latter may be called Occasional, or Accidental, Residents. They ſeldom continue with us except in open, mild winters. Theſe, in the preceding liſt, are marked with an aſteriſk.* I have had the leſs heſitation in blending them together, becauſe it is highly probable, that in a

> tract of country ſo extenſive and various as the ſtate of Pennſlvania, ſome of the birds, which about Philadelphia are only occaſionally reſident, may be more generally ſo in other parts of the country. (Barton, 1799, p. 13)

Though his work was not widely read, it is important in ornithological discourse because it represents a very early systematic method of field observation, presented in a tabular form reminiscent of Renaissance writer Petrus Ramus (Tebeaux, 1991a). Botanist Barton used his botanical expertise to delineate the relationship between bird migration and the seasonal changes in vegetation, topics of continued importance to ornithology throughout its history.

More significant to the development of scientific discourse in American ornithology were works appearing in the nineteenth century. A key figure in the placement of American ornithology in a scientific framework was Napoleon's nephew Charles Lucien Bonaparte (1825–1833), who supplemented Wilson's work, adding 60 species to his compilation. Bonaparte's foundation in European ornithology brought the topic of systematics to the attention of American ornithology (Coues, 1903, p. xx) with his "Observations on the Nomenclature of Wilson's Ornithology," a series published in the journal of the Philadelphia Academy of Natural Science in 1824–1825. He also produced a number of catalogues of birds of the United States between 1827 and 1838. Birds' ranges, descriptions, and behavior were detailed and authoritative, evidence of Bonaparte's scientific orientation.

Such interest in the ordering of bird families prompted even Audubon, whose color plates had already assured his popularity among naturalists, to offer two works arranged systematically. One was a less expensive octavo edition (1840–1844), with 500 plates and accompanying text rearranged systematically for each bird. His *Synopsis of the Birds of North America* (1839) afforded

> a methodical catalogue of all the species hitherto discovered . . . [and presented] an arrangement of these birds, so characterized, that a person desirous of studying them, might, without much difficulty, be enabled to discover their names, and trace some of the most important features in their organization. (p. v)

This increasing interest in the classification of birds indicates the beginnings of specialization in ornithology, a sign that the science had begun to professionalize. However, theoretical interest in ornithology as evidenced by the study of systematics and taxonomy did not transfer to the university curriculum as a specific program of study in ornithology until the twentieth century. Thus, though nineteenth-century university science courses, such as those at Harvard, might have instilled in their students the scientific method, young ornithologists would not have had the opportunity to study birds extensively. As a consequence, nineteenth-century ornithological discourse did not emerge from the university, but rather from the country's natural history and science museums, where bird study skins were preserved for scientific study.

During the first half of the nineteenth century, museums in Philadelphia and Washington, D.C., were the centers of ornithological study in America (Mayr, 1951/1975, p. 366). By 1856, the bird department of the Academy of Natural Sciences founded in Philadelphia in 1812 had become "the largest and richest in the world, with 29,000 specimens (23,000 in glass cases)" (Stresemann, 1951/1975, p. 243). The museum's massive collection prompted P. L. Sclater (1857) to comment to the Zoological Society of London that the Philadelphia Academy's "department of Ornithology . . . is probably superior to every Museum in Europe, and therefore the most perfect in existence" (p. 1). It was easily the greatest bird collection in the world, since the Smithsonian collection did not begin competing with Philadelphia's until the arrival of specimens from the American West. Not until after mid century did universities and other cities begin to establish museums containing ornithological collections (Mayr, 1951/1975, p. 366), though there can be noted a few exceptions, like the University of Michigan's collection, begun in 1837, and the Boston Society of Natural History accessions in 1830 (Peters, 1933/1974, pp. 137, 140). Harvard's bird skin collection did not begin until 1847 (p. 133), though their Museum of Comparative Zoology, established in 1859 by Louis Agassiz, was one of the earlier university museums. Amateurs such as Henry David Thoreau were among those who contributed ornithological specimens.

BEGINNINGS OF PROFESSIONALISM IN AMERICAN ORNITHOLOGY

A scientific community's growing sense of itself can take many forms. There must first be a body of knowledge to be manipulated and expanded upon. The previous pages have demonstrated two such bodies of knowledge in natural history and systematics. There must likewise be an enthusiastic group of aficionados interested in devoting their working lives to expanding this knowledge. By the 1850s there were two camps of devotees—the general public interested in the natural history of birds and scientists interested in the technical aspects of ornithology. Third, there must be financial support for such an endeavor. In the mid nineteenth century, the federal government and museums provided this support.

A discourse indicator of professionalization is the appearance of reference texts on the subject. The discourse of American ornithology at this time began producing these texts, thus reflecting this growing sense of professionalism, which would lead ultimately to the founding of a professional ornithological organization in the United States. As early as 1840, Thomas Brewer recognized the developing scientific community of American ornithologists. Significant here as a founding member of the first ornithological club in America, Brewer offered at that time an edited edition of *Wilson's American Ornithology*. Its authority as a reference work rests on its extensive descriptions of birds; notes by a renowned British ornithologist William Jardine; and most importantly a "synopsis" of American birds, a taxonomic classification including current scientific and common bird names, as for example this entry from the synopsis of the Harlan's Hawk:

> 15. Falco Harlani.—Black Warrior.
>
> Plumage, chocolate brown, glossed with bluish gray; tail, lighter than back, narrowly barred with brownish black, and tipped with brownish red; lower parts, paler, anteriorly streaked, posteriorly barred with brownish black. *Male*, 21,45. *Female*, 22. Habitat, Louisiana.
>
> Falco Harlani, *Aud.* i. 441; v. 380; plate 86. *Nutt.* i. 105.—Buteo Harlani, *Aud. Syn.* p. 7. (Wilson, 1808-1814/1840, p. 684)

Another indication of approaching professionalization was Thomas Nuttall's *A Manual of the Ornithology of the United States and of Canada* (1832; 1834). Of significance was his inclusion of a 30-page general introduction to ornithology, which suggests that Nuttall recognized the need to compile in one place the rapidly increasing body of information about birds.

The appearance of an ornithological textbook in America implies the presence of an audience of professionals and bird students. Such a work was *Key to North American Birds* (1872) by Elliott Coues, a principal figure in the transforming of American ornithology into a science. His key was a landmark in American ornithological discourse, serving in multiple capacity as an ornithological history and text, field manual, updated taxonomy, and identification key, ranking "among the most influential of ornithological publications" (Cannell, 1983, p. 355). Coues' thoroughness is evident from the following excerpt, which not only describes the bird but also provides cross-references of sources:

> **A. brewsteri ?** (To Wm. Brewster of Cambridge.) Brewster's Linnet. With the general appearance of an immature *A. linaria*, this bird will be recognized by absence of crimson on crown, no black throat-spot, a sulphur-yellowish shade on lower back, and somewhat different proportions. Wing 3.00; tail 2.50; tarsus 0.50. Waltham, Mass., Nov. 1, 1870; one specimen known. *Ægiothus flavirostris*, var. *brewsterii* Ridgw. Am. Nat. vi, July, 1872, p. 433; Hist. N. A. B. i, 1874, p. 501; *Acanthis brewsterii* Ridgw. Man. 1887, p. 398; A. O. U. Hypothetical List, 1895, p. 330, No. 17. See Brewster, Bull. Nutt. Orn. Club, vi, 1881, p. 225. Conjectured to be *Acanthis linaria X Spinus pinus*. (Vol. 1, p. 391)

In addition, his *Field Ornithology* (1874) represented an early attempt to standardize methodology and taxonomy in America. The following excerpts from the Table of Contents show the breath of his manual:

Field Ornithology
- Implements for Collecting, and their Use
- Various Suggestions and Directions for Field-work
- Hygiene of Collectorship
- Registration and Labelling
- Instruments, Materials, and Fixtures for preparing Birdskins
- How to make a Birdskin
- Collection of Nests and Eggs
- Care of a Collection

General Ornithology
- Definition of Birds
- Principles and Practice of Classification
- Definitions and Descriptions of the Exterior Parts of Birds
- An Introduction to the Anatomy of Birds (n.p.)

Bird lists, once simple compilations in state histories and travel narratives, became more sophisticated in the nineteenth century, as publications describing the birds of individual states appeared, as for example De Kay and Giraud for New York (1842–1844), in addition to catalogues of birds for numerous New England and Midwestern states (Coues, 1903, pp. xxiii–xxv). Such activity prompted ornithologist Elliott Coues (1871) to write of the East: "Our local lists of the Atlantic coast may now be considered nearly complete, and they collectively afford a very perfect ornithological map from Labrador to Florida" (p. 365).

In 1862, ornithology had found a place for itself, at least temporarily, in the Department of Agriculture. The beginning of economic ornithology appeared as the USDA established the organic act, which included a clause concerning the introduction and protection of birds beneficial to agriculture, yet another indication of increasing professionalization through specialization. Though later appropriation bills omitted the clause, research and published reports in the 1860s acknowledged the value of birds to the farmer (McAtee, 1933/1974, pp. 113–114), providing further evidence of the practical value of ornithology.

The most significant association between government and ornithology, and thus a concerted boost to the professionalization of American ornithology, was the gathering of ornithological data during the U. S. Army's surveys for railroad routes to the Pacific Ocean during the mid nineteenth century. As assistant secretary of the Smithsonian Institute in 1850, Spencer Fullerton Baird, avidly interested in ornithology, was able to advance the study of birds in conjunction with government surveys of western territory, thereby making the first significant American contribution to ornithological discourse. In order to assure that scientifically valid data were collected, Baird selected a team of Army medics willing to serve him by recording ornithological data and collecting and preserving bird specimens. Unlike the self-taught amateurs or artists before them, these medics knew scientific method, taxonomy, and anatomy. They could consequently provide accurate, detailed ornithological records of American birds as no one had done before. Beginning in 1853, U.S. Army expeditions began surveying four main routes from the Mississippi River to the Pacific Ocean (Hume, 1942, p. 3). Their subsequent reports, published by the War Department between 1853 and 1856, "mark the emergence of a sophisticated American science dependent on Europe only for the basic techniques of classification" (Goetzmann, 1959, p. 331). Volume 9 of *Reports of Explorations and Surveys* (1853–1856), which was Baird's report on the birds his medics had collected, is recognized as a landmark in American ornithology. In 1858, the Smithsonian issued a shortened version by Baird, *Catalogue of North American Birds* (1858), at that time the most complete, systematic, nomenclaturally-sound listing of North American birds. It began with tables of orders and suborders of birds followed by short descriptions of North American birds, cross-referenced to page numbers in the *Report*. The following excerpt shows the depth of coverage:

> 170. *Geothylpis trichas*, (Linn.) Cab. Maryland Yellow-throat. North America from Atlantic to Pacific 241
> (171.) *Geothylpis velatus*, (Vieill.) Cab. Green-headed warbler. West Indes and South America. 243
> 172. *Geothylpis philadelphia*, (Wils.) Baird. Mourning Warbler. Eastern North America as far west as Independence, Mo 243
> 173. *Geothylpis macgillivrayi*, (Aud.) Baird. Macgillivray's Warbler. Pacific coast of North America, south to Gulf of California, and across to Monterey, Mexico. In Rocky Mountains to Fort Laramie 244
> (p. xxxii)

It has become the "basis for all the guides and check lists" (Kastner, 1986, p. 17) used today.

As a result of these Army surveys, which occurred periodically throughout the remainder of the century, many of these medics published guides and handbooks based on their ornithological work (Hume, 1942). With the exception of John Cassin's *Illustrations of the Birds of California* (1856), most of the survey-generated texts, the fruits of Army medics' work, were published in the 1870s. For instance, there was another California guide in 1870; an 1875 report of the birds of the southwestern United States; Coues' publications on the birds of the northwest in 1874 and of the Colorado

Valley in 1878, both based on U. S. Geological Surveys; and *History of North American Birds* (1875), by Baird, Thomas Brewer, and Robert Ridgway.

Baird and his army medics provided the prestige needed to give credibility to the fledgling American professionals because, through their acceptance of Darwin's theory of evolution, they were able to incorporate their observations of variation in western American birds into a truly systematic nomenclature, while European ornithologists were still struggling with both Darwin's theories and ornithological systematics. Now, the time was ripe for an American ornithological society.

THE INFLUENCE OF BRITISH ORNITHOLOGY

Before turning attention to a strictly American ornithology, a brief look at key figures in the development of British ornithology will give insight into the major concepts and concerns discussed in this chapter, which have found their way into American ornithology. Elliott Coues (1903), prominent late nineteenth-century American ornithologist and historian, has written that several British ornithologists prepared nineteenth-century American ornithology for its great accomplishments (p. xvi). Besides two previously noted eighteenth-century ornithologists Mark Catesby and William Bartram, Coues singled out four nineteenth-century British figures for their noteworthy influence: John Latham, Thomas Pennant, George Edwards, and William Macgillivray. These figures represent Britain's major contribution to the evolution of American ornithology.

Due to his reputation as the leading British ornithologist of the late eighteenth and early nineteenth centuries, John Latham is a fitting transition figure between these two centuries. He is known for two major ornithological works, both taxonomic handbooks. His three-volume *A General Synopsis of Birds* (1785) contained 106 plates produced by Latham himself. Based on physical characteristics, the work was divided into land and water birds, a division used previously by John Ray. He also contributed an 11-volume work, *A General History of Birds* (1821–1828), containing 193 of Latham's own color plates. Not only was he the first to introduce 24 American birds, he was more important in ornithological discourse because, in his attempt to update European ornithologists Brisson and Buffon, Latham produced the most complete works of his time. Most significantly, he incorporated into the already established body of ornithological knowledge recent discoveries made by exploratory expeditions around the world during the eighteenth century. Having incorporated findings from these expeditions into his ornithology, Latham can be seen as the bridge, as Elsa Allen (1951) described him, between more simple ornithological studies stemming from exploration of new lands and advanced compilations of facts from around the world "at a time when zoological data were increasing with almost modern speed" (p. 496).

The second influential British ornithologist, Thomas Pennant, prepared *Genera of Birds* (1773) for a natural history professor as an elementary text on birds, a "forerunner of the modern laboratory notebook for the study of birds" (E.G. Allen, 1951, p. 492). It was reissued with 15 plates in 1781. His most important work was *Arctic Zoology* (1785), specifically the second volume devoted solely to Arctic birds; 427 of these are American birds with 15 accompanying plates. He skillfully compiled observations by numerous travelers and correspondents, including the Bartrams, as well as by the standard ornithological works of his time. Thus, his completeness is noteworthy though his accuracy is sometimes lacking. Although he opted for vernacular over scientific bird names, his descriptions were later used by taxonomists as the source of many binomial names (E.G. Allen, p. 492). He was also one of four late eighteenth-century writers of picture books of birds. His *British Zoology* (1768–1770) had run through four editions by 1777 (Stresemann, 1951/1975, p. 296).

The key British figure of the century was George Edwards, considered the father of British ornithology. His works included plates of birds with accompanying descriptions and the location of sighting. His four-volume *Natural History of Birds* (1743–1751) contained 249 plates of birds, engraved by Edwards himself, who also colored most of them. He included 57 American species, showing the European interest in American birds. Because of this American emphasis, Coues (1903) considered Edwards to be a key figure in eighteenth-century American ornithological history (p. xiv). His other major work *Gleanings of Natural History* (1758–1764) incorporated further accounts of birds with those of mammals. Of special significance was a complete life history of an American bird, the Ruffed Grouse, about which Edwards had received notes from John Bartram, a regular correspondent. Edwards' compilation had come from a paper on the subject delivered in 1754 to the Royal Society, for whom he had presented a paper on another American bird, the Red Phalarope, also in the 1750s. Edwards was also known for publishing the 1771 edition of Mark Catesby's colonial natural history.

William Macgillivray, another acknowledged British ornithologist of the time, published his most significant continental work, *Descriptions of the Rapacious Birds of Great Britain*, in 1836. Macgillivray influenced American ornithology, however, with his editorship of Audubon's *Ornithological Biography*, adding "accurate morphological (and partly anatomical) details" (Stresemann, 1951/1975, p. 314) and in the process lending scientific credibility to Audubon's work. Coues (1903) attributed to Macgillivray's contribution the first systematic presentation of the anatomical structure of American birds (p. xxii).

There were, of course, many other British publications during the evolution of American ornithology, though their subject matter was primarily natural history. On the other hand, British colonial accounts may be said to have influenced the discourse of American ornithology in the same way that colonial America in general relied on its European counterparts—that is, as models for imitation.

In the seventeenth and eighteenth centuries, accounts of expeditions to Hudson Bay in search of a northwest passage also detailed the bird life of the region. Accounts by members of the Hudson Bay Company included those by Humphrey Martin and Thomas Hutchins. Of special note in Hudson Bay accounts was Samuel Hearne's *A Journey from Prince of Wales's Fort in Hudson's Bay to the Northern Ocean* (1795) for its careful descriptions of some 53 species of North American Arctic birds. Captain James Cook's voyages around the world in the 1770s also increased ornithological knowledge. On Cook's first voyage was artist Sydney Parkinson, whose 900+ drawings and journal contributed to the ornithological record. Whereas ornithologists J. R. Forster and his son Georg accompanied Cook's second voyage, on his third voyage was artist William W. Ellis, who contributed to the ornithology of the Pacific. Finally, an expedition to Labrador produced a work by Captain George Cartwright, *A Journal of Transactions and Events during a Residence of nearly Sixteen Years on the Coast of Labrador* (1792), with ornithological observations scattered throughout.

Two strictly natural history publications of the eighteenth century are noteworthy. *Natural History of Birds*, published in 1731 by Eleazar Albin, was important for its account of external parasites of birds, rather than for bird description per se. In addition, Gilbert White's *The Natural History and Antiquities of Selborne* (1789) documented the value of careful field observation of the living bird.

Eighteenth-century travel literature also contributed to the accumulating discourse of American ornithology. An early work of the century was John Lawson's *The History of Carolina*, which contained accounts of 56 land birds, at times intermingling fact with superstition, but most notably giving distinguishing characteristics of each bird. Its popularity was attested to by its three English editions between 1709 and 1722. A less accurate work on similar lines appeared in 1737 by John Brickell, *The Natural History of North-Carolina*, which included crude plates of birds. Travel narratives by foreign writers also added to the ornithological natural history record. Two such works were those by Frenchman Jean Bossu (*Travels through That Part of North America Formerly Called Louisiana*) and Swede Peter Kalm (*Travels into North America*), both translated into English and published around 1771. Kalm, for instance, gave life history accounts of the hummingbird and barn swallow, in addition to annotated lists and descriptive accounts.

In addition to books on British oology, Thomas Bewick's *The History of British Birds* (1797–1804), was a popular nineteenth-century two-volume work, emphasizing natural history accounts over the classification of birds. Illustrated in woodcuts, his work is significant because it helped to popularize the study of birds by enabling amateurs to pursue field study with the help of his woodcuts (Farber, 1982, p. 29). Bewick's own comments about his writings ring a transcendental tone: "My object was to inveigle youth onward by the vignettes to the study of natural history, by which they would be led, in contemplating the works of nature, up to nature's God" (as cited in Farber, pp. 28–29). Two other British works, by Lord Edward Stanley (1835) and ornithologist William Yarrell (1839–1843), were based on these authors' extensive personal bird collections. Yarrell's handbook became the standard British reference of the time (Farber, p. 63). Bechstein's *Cage and Chamber-Birds* (1877) detailed his rigorous classification of birds, though his inclusion of natural history accounts showed this genre's persistence as part of ornithological discourse (Stresemann, 1951/1975, pp. 293–294).

In general, however, British ornithology had only limited influence upon its American counterpart because American ornithologists during their emergence as a unified body in the latter half of the nineteenth century quickly accepted Darwinism, subsequently changing their taxonomies and consequently their research priorities to accommodate evolutionary theory. A similar decision was not made by British ornithology until the early twentieth century (see, for instance, Snow, 1992, p. 87), causing nineteenth-century ornithology on the two continents, and in turn their discourse, to diverge in very different directions.

THE BIRTH OF AN AMERICAN ORNITHOLOGISTS' UNION

Cambridge, Massachusetts, may rightly be called the birthplace of a strictly American approach to ornithology. By the 1860s, the city had become a center of ornithological activity, producing a number of highly qualified future ornithologists (Mayr, 1951/1975, p. 369). Two high school boys, William Brewster and Henry Henshaw, began organizing meetings, which in 1873 became the Nuttall Ornithological Club, the first organized bird club in America. An offshoot of the club was the *Bulletin of the Nuttall Ornithological Club*, begun in 1876, the first periodical in America devoted strictly to ornithology. Then, in 1883, partly inspired by Cambridge's Brewster, was formed the American Ornithologists' Union and its professional journal, *The Auk*, which had evolved from the *Bulletin*[5], and which is the focus of the present study.

Despite this initial attempt at professionalization, the intermingling of the amateur and professional continued, because ornithological fieldwork relied not on elaborate scientific instrumentation, but on only one instrument, the gun, the primary field instrument until the twentieth century, when effective prism binoculars became available for general use. Thus, observation and collection in the field remained the primary method of bird study both for the amateur and professional. It is not until ornithologists' retreat to the laboratory during the mid twentieth century, as described in chapter 4, that technology made possible sophisticated analyses such as electrophoresis for DNA research, radar for bird migration analysis, sound spectography for bird song, and inferential statistical software packages for field-data analysis, effectively delineating the distinction between amateur and professional and thereby greatly curtailing amateur influence in ornithological science.

But how really different is amateur and professional discourse? Are there interconnecting links between both forms of writing? Two circumstances appear possible. Perhaps changes in the continuum from natural history to experimental science discourse are specific shifts from one audience or situation to another. Or these changes may be gradual, represented by boundary-creating linguistic changes. One value of such a study of ornithological discourse is to search for general models of natural history and experimental science discourse. Consequently, the next chapter will suggest models describing the discourse of the amateur and professional ornithologist. It will then delineate these two poles of ornithological discourse further through case studies of the writings of two groups of two ornithologist-authors, who represent amateur natural history and experimental scientific discourse, respectively.

How was the discourse of American ornithology affected by the intermingling of amateur and professional? What was the interconnection between natural history and systematics, the two traditions of bird study? Was the fledgling American Ornithologists' Union of 1884 able to divorce itself quickly from amateur influence? In chapter 4, the study of the progression of research topics and methodology by American ornithology clarifies the relationship by showing the interplay between natural history and experimental science discourse during American ornithology's development from 1884 to 1990. The chapter will also use the case study method to analyze the writings of ornithologist Frank M. Chapman, whose participation in both amateur and professional ornithology gave him expertise in writing both natural history and experimental science discourse. Thus, his writing should exemplify a transitional discourse—a bridge from one discourse type to the other. Delineating the social, linguistic, and structural differences among these three models will clarify the forces of change demonstrated by the historical overviews of ornithological discourse in this chapter and in chapter 4. Describing the interplay between natural history and experimental science discourse, between amateur and professional, will give context and perspective to the study presented in chapter 5 of American ornithology's principal journal, *The Auk,* since the continuum of change in these writers' discourse should parallel the professionalization of American ornithology.

3

Discourse Models for Natural History and Experimental Science

In his analysis of European ornithology's emergence as a scientific discipline between 1760 and 1850, Paul Lawrence Farber (1982) focused on the writings of two eighteenth-century French ornithologists: Georges-Louis Leclerc de Buffon and Mathurin Jacques Brisson. *Histoire Naturelle des Oiseaux* (1778–1786) by Buffon was one of two major eighteenth-century treatises on birds. It was based on nine volumes on birds from his 44-volume *Histoire Naturelle*. Of special significance were his series of color plates known as the "Planches Enluminées." Buffon extended the study of birds from a mere classification of specimens to a more expansive interrelationship of birds with their natural environment. By 1800, his works had been translated into all major European languages and had even been shortened into children's versions (Farber, 1982, pp. 68–69). Since he focused on birds' habits and offered "short cuts in bird description" (E.G. Allen, 1951, p. 500), the work foreshadowed the popularization of birds, which would occur in the next century. Earlier, in 1760, Mathurin Jacques Brisson published *Ornithologie*, the other major eighteenth-century treatise on birds. It was a Latin-French text, which developed a taxonomic reclassification of birds based on specimen collections. As Elsa Allen (1951) has pointed out, the work served as a "model of method in observation and description to modern ornithologists" (p. 499).

The works of Buffon and Brisson are significant because they propose to provide a comprehensive treatment of birds. Farber characterized the two figures as symbols of the ongoing question that until the early twentieth century fueled ornithology's main branch of study, systematics: Which has theoretical priority, natural history (Buffon) or morphology (Brisson)?

These two influential, popular texts helped to establish the two main camps into which the study of American ornithology has fallen. Leclerc de Buffon was the "leading natural historian of the second half of the eighteenth century," whose writings helped to popularize science during the Enlightenment (Farber, 1982, pp. 15–17). Though Buffon also used a collection extensively (that of the King's *Cabinet du Roí*, later to become the founding collection for the French national museum), he believed that an understanding of the natural history of the bird had precedence, before classification was feasible. His ornithological work, *Histoire Naturelle des Oiseaux* (1778–1786), another effort at an encyclopedic account after the manner of Renaissance writers Aldrovandi and Gesner, had a literary bent, since Buffon was more of a stylist than Brisson. As a result, he strived to provide his readers with pared down, easier-to-understand descriptions of birds, with shortcuts to identification (E.G. Allen, 1951, p. 500). He surveyed the mass of knowledge presented about birds and transformed it into more general laws of nature. That is, he sacrificed technical accuracy for readability and usability.

In contrast, Brisson, as Buffon's counterpart, oversaw an extensive private natural history collection of French collector René-Antoine Ferchault de Réaumur. As a result, he had access to a superb private collection of bird skins—in fact, the largest collection in Europe. To Brisson, bird study meant the classification of these skins. Thus, his major publication, *Ornithologie* (1760), was like "a museum curator's relationship to a large private collection" (Farber, 1982, p. 10), and contained his major contribution to ornithology, the arrangement of birds in 26 orders according to beaks and claws. This concept, of course, was not new; Francis Willughby and John Ray had already published a similar account in England in the seventeenth century. But theirs was the result, at least in part, of personal trips around Europe observing and documenting birds. Brisson, on the other hand, had the opportunity to observe large numbers of specimens in the hand, a luxury not afforded Ray and Willughby, who lived before the discovery of effective methods to preserve fragile bird skins.

Farber (1982) considered these two distinct approaches to ornithology as "models" (p. 26) that helped to delineate ornithology as the scientific study of birds, rather than a mere collection of diverse texts such as cookbooks and encyclopedias (p. 79). As evidence of the acceptance of these implicit definitions of ornithology published by Buffon and Brisson, Farber cited the fact that there was in European ornithological literature between 1780 and 1830 no significant discussion of domestic fowl breeding, parlor bird observation, or anatomical and physiological comparisons (p. 47). The emphasis, instead, was on natural history and specimen study.

DEVELOPING THE DISCOURSE MODELS

Following Farber's analysis, Buffon and Brisson may be considered representatives of the two traditions that should be the focus of a study of American ornithological discourse: the *naturalist*, who seeks a complete natural history of birds; and the *taxonomist*, who deals with nomenclature and systematics[1]. From these two traditions, we can develop two distinct discourse models for American ornithology. The naturalist tradition represents what will be called here the "natural history model," the primary model used by the amateur. On the other hand, the taxonomist represents the "experimental science model," which is employed by empirical science and, in its most mature form, by theoretical science. It is the discourse model used by professional science. Though the professional scientist may employ natural history discourse in his or her writing, the amateur may not—and, given the state of contemporary professional science, will likely not—employ experimental science discourse. The theoretical sophistication inherent in professional science will ultimately restrict, and often exclude, participation by the amateur.

The Continuum of Scientific Perspectives on Nature proposed by Killingsworth and Palmer (1992) (see Figure 3.1) clarifies the relationship between natural history and experimental science being used here. The continuum evolved from interviews with ecologists at Memphis State University to describe the progression from natural history study to experimental scientific research. Early in the progression toward scientific research, elementary field observations give way to more quantitative approaches, first by the collection of basic measurements, followed by initial attempts at experimentation through simple manipulation of the environment. As a means for imposing more control and/or making more precise or intricate measurements, the laboratory becomes the center for research. Given this controlled environment, sophisticated technology may now be used to solve theoretical, as well as complex empirical, questions. Thus, in practice, natural history study gives way to increasingly empirical approaches, leading to theoretical experimentation. Correspondingly, natural history discourse gives way to experimental science discourse; the amateur gives way to the professional. The study to be described in chapter 4 of topics and methodologies in *The Auk*, American ornithology's major professional journal, over the past century reveals a similar progression by American ornithology.

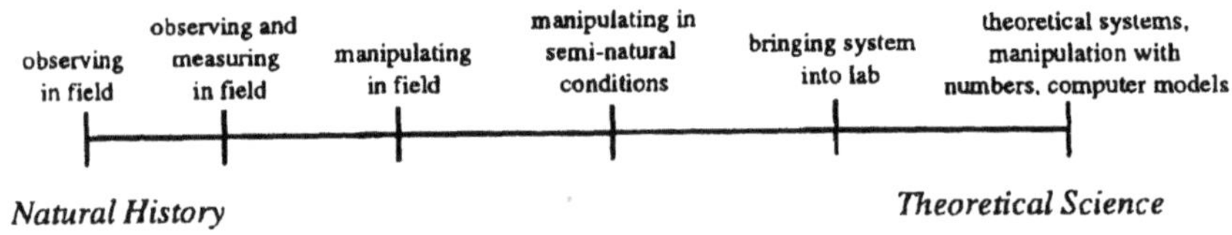

FIGURE 3.1. Continuum of Scientific Perspectives on Nature (Killingsworth & Palmer, 1992, p. 112). Copyright © 1992 by the Board of Trustees, Southern Illinois University.

In developing these two models of discourse (natural history and experimental science), the following three aspects of discourse will be described:

- rhetorical considerations: persona, audience, and context
- linguistic features
- format and physical appearance

In addition, case studies of four writers—two natural history and two experimental science writers—will complement this analysis of their respective discourse. The ultimate goal is to establish the natural history and experimental science discourse models as the two extremes of discourse used by American ornithologists during their century of evolution toward professionalization. This study will then employ these two models as benchmarks against which the evolution of professional ornithological discourse in America can be compared. In other words, these models represent ideals that are not absolutely fixed. For instance, in practice, scientific papers may represent trends in one direction or the other, rather than appearing as perfect examples of one discourse model or the other. Thus, by identifying the extremes of discourse, it will be possible to study the articles published in the past 100 years of *The Auk* by comparing the characteristic features of these articles—as intermediate forms of discourse—with the features identified by the two discourse models, thereby documenting the slow progression from natural history to experimental science. A "model" then represents here a generic idea, or paradigm, which is to be tested or against which actual examples are to be contrasted and evaluated.

It must be emphasized here that this study attempts to identify the major discourse *models* present in American ornithology. These models should not necessarily be equated with scientific *approaches* to the study of ornithology. Indeed, there have been, and continue to be, many approaches to ornithological study (see Brush & Clark, 1983), each of which may have its own unique discursive subcategories.

Rhetorical Considerations in Developing the Discourse Models

An effective discourse model clarifies not only the writer's own persona, but also that of his or her audience. It describes the relationship between writer and audience, as well as the mechanisms that allow the two to interact. It offers a broader view of the social, philosophical, and intellectual milieu within which the writer operates. Finally, it shows how the discourse being defined is distinct from discourse with which it competes. In his discussion of "forums," the approved channels of communication within discourse communities, James Porter (1986, pp. 46–47) has offered a series of questions which delineate discourse conventions within these forums. Using these questions as a guide for developing discourse models, this study offers a set of five questions which provide a basis for articulating the major rhetorical concepts essential for each discourse model:

1. *Who am I and for whom am I speaking?* Audience, of course, is of prime importance in any discussion of discourse. The development of writers' own personae (that is, how they see themselves) is pivotal in determining how they view their audiences, whom they must see in relationship to themselves. What I say to others is determined in part by how I want to be perceived by them. What I say is also determined by my status in that discourse community—that is, by the kind of privileges I have in speaking. Thus, an accurate discourse model must describe how writers of the discourse perceive both themselves and their audiences and how their audiences perceive them in return.
2. *What forces bind discourse partners together?* It is not sufficient merely to consider writers' views of themselves. Since audience shapes most discourse, it is imperative to consider the social context in which a writer writes. These sociological considerations must include the roles of power and authority in the formulation of discourse (Hodge & Kress, 1988).
3. *What is their worldview?* How writers see themselves in the scheme of things also affects their discourse. What writers say and how they say it are determined in part by the moral and social norms they profess, their philosophical and theological outlooks, their perception of their own relationship to other cultures and societies and to other professions and organizations, and the importance they attach to their own writing as part of the broader discourse in their field. According to George Lakoff and Mark Johnson (1980), language (i.e., discourse) is pivotal to a person's acquisition of a worldview. In their analysis, metaphor in part determines worldview and social reality since each culture must define its social reality, which affects a culture's physical reality; that is, what is real for individuals consists of their social reality

and the way this reality shapes their experience of the physical world (p. 146).

4. *What is the status of their knowledge and how are knowledge claims made?* The process by which discourse concretizes and formalizes knowledge is influenced by the status of that knowledge. That is, the more established, recognized avenues there are for argument, the less discourse must be manipulated. In turn, the more knowledge claims must be argued, the more rhetoric comes into play. In short, though discourse may affect knowledge, the status of knowledge likewise affects discourse.
5. *How is the same knowledge used by both discourse groups?* How amateurs and professionals plan to use the knowledge they generate affects the form the discourse takes. Though both types of writers function by manipulating the same knowledge base, there are varying levels of sophistication at work in the process. Thus, the form and function of the discourse of both groups, though it may be on similar topics, will diverge and thus be a characteristic means of distinguishing discourse models.

Linguistic Features of Natural History and Experimental Science Discourse

In this chapter, the linguistic characteristics of both natural history and experimental science discourse models are described by studying the writing styles of four authors whose writing is representative of natural history and experimental science discourse. Analyzed are two samples each of the writings of Henry David Thoreau and John Burroughs for the natural history model, and two samples each of two contemporary ornithologists, named here A and B, for the experimental science model. As the case studies show, these authors are representative of each discourse model. Thoreau and Burroughs were not only recognized as natural historians in their own right, but their observations were acknowledged by American ornithology as well. Ornithologists A and B are systematic ornithologists who conduct DNA research, a topic that is not only on the "cutting edge" of contemporary ornithological research but also a continuation of the taxonomic research that has evolved throughout ornithological history, as has been shown in chapter 2.

It must be recognized, however, that it is impossible to separate completely an individual writer's influence from any piece of discourse. Individual writers make decisions about their writing based upon their own past experiences as well as upon objective norms established by the discourse and by the audience itself. Thus, there will always be some variance in the empirical results of such a study since individual variation is inherent in all writing. It should, therefore, be the object of such a study to look at the empirical data gleaned from individual authors' styles as *indications of* and *generalizations about* the type of writing style to be expected from a particular discourse model. In particular, it is recognized that the samples used to develop the linguistic components of these two models are small. However, here we are merely looking for benchmark values to show the extremes of American ornithologists' discourse. Chapter 4 will propose a transition model within which most ornithological discourse falls. Chapter 5 will refine this model, and chapter 6 will interpret the three models in light of the past 100 years of American ornithologists' discourse.

Given these limitations, several linguistic features can be used in describing discourse models for natural history and experimental science—that is, for amateur and professional discourse in American ornithology, respectively. Linguistic features—such as diction choices and variations in sentence length and complexity—indicate audience perspectives, sophistication of knowledge, and means of exclusion. Bazerman (1988) has offered an excellent methodology for analyzing these features of discourse. As linguistic markers for making factual claims, the use of evidentials (Chafe, 1986) is another linguistic feature that allows listeners to evaluate users' attitudes about and confidence in the information being conveyed.[2] Lakoff and Johnson (1980) have demonstrated the importance of identifying the major tropes, metaphors, and figures of speech inherent in all language choice. These language features identify users' worldviews, philosophies, and social and cultural perspectives.

A sentence-level analysis[3] focuses on the writer's use of the following devices:

- Personal pronouns, passive voice, nominalization, and metadiscourse. Occurrences of these were counted and variations in use noted by means of a frequency distribution table.
- Technical terms. The presence of technical terminology and the amount and frequency of the use of definitions were noted and, where possible, related to their historical context.
- Average sentence length. Frequency distribution was determined.
- Sentence complexity. Sentence complexity was shown by the occurrence of topics in subject position, subjects as place holders, use of complex clauses, and interruptions of subject and verb.
- Evidentials. Expressions indicative of reliability of evidence, belief, induction, hearsay, citations, deduction, hedging, sensory evidence, and expectations (Chafe, 1986) measure the extent to which discourse can be categorized as professional.

The major hypotheses in this comparative study of the progression from natural history to experimental science discourse are that

- sentence lengths will increase
- metadiscourse will increase in variety
- nominalizations and technical terminology will increase, whereas personal pronouns and definitions will decrease
- sentences will become more dense, whereas clauses will increase in complexity, emphasizing cause-effect relationships
- passive voice will increase
- argumentation will be more empirically based, focusing on the use of citation, induction, and assertions of reliability

The study described in this chapter verifies most of these hypotheses, though sentence length and complexity and the use of metadiscourse and definitions do not follow expected patterns. Nor is inductive discourse used significantly by the scientists and naturalists sampled. In chapter 5, the same analysis is applied to a sample of 44 articles from a single professional journal, *The Auk*. Chapter 6 discusses the results of the study, including the reasons for inconsistencies and unexpected findings.

Format and Physical Appearance of Discourse

A description of the physical appearance and structure of the discourse delineates further the variations in discourse models. Central components in this part of the discourse model are (a) the available genres, (b) the use of visuals, and (c) document formatting with particular attention to the frequency and use of headings, IMRaD formats, and the presence of Acknowledgments and Literature Review sections. The presence of these features of discourse indicates social, epistemological, and contextual pressures influencing the discourse, that is, as measures of professionalization (see, for instance, Gross, 1990).

NATURAL HISTORY DISCOURSE MODEL

In analyzing natural history discourse, this study employs the definition of natural history proposed by Myers (1990): "Natural history gives a written account of actions of particular animals at a particular place or time, recorded by particular observers" (p. 195). Authors of these texts may be identified as those who use past tense; describe setting gratuitously; characterize, if not anthropomorphize, animals; and focus on the observer's perspective (Myers, pp. 195–203). In ornithology, such texts are often associated with personal bird lists; descriptions of the setting, including time, place, and habitat where a bird was observed; historical accounts of the occurrence of a species; physical, though not taxonomic, descriptions of a bird; and extended descriptions of a bird's behavior using common, rather than specialized, vocabulary. Given this definition, described below are the essential premises upon which natural history discourse is based, followed by the model's typical linguistic and structural features.

Rhetorical Considerations

First, answers to the following five questions will clarify the context within which natural history discourse occurs.

> *Who Am I and For Whom Am I Speaking?* Of course, anyone interested in birds may be called an amateur. A person may choose to study birds for a variety of reasons: as a hobby, for relaxation, for data collection, and even as a competitive sport. As a general rule, however, amateurs are preoccupied with acquiring personal knowledge about birds and their lives. Amateurs may wish to observe a bird or behavior they have read about in an authoritative source or hope to gain some insight or make some observation previously unreported. The amateur endeavor is often not a solitary affair, since many amateurs join bird or nature clubs, thereby becoming one of many observers of birds. This companionship, which provides a forum for amateurs to tell each other what they have seen, substitutes in a limited way for the professional forums and collegiality readily available to professionals. Similarly, those amateurs who choose to write about their experiences speak to "everyone" figures, a generalized, lay audience like themselves and their bird club associates, to whom they can impart some knowledge gained from personal insight. Their forum is ordinarily the newsletters or periodical magazines published by bird clubs or nature or conservation societies. State wildlife and conservation agencies also have non-refereed periodicals to which amateurs also contribute their ornithological observations and reflections.
>
> There are likewise some instances in which amateurs may believe that they have a contribution to make to "science." In such circumstances, they may attempt to overstep the boundary between amateur and scientific audiences. Amateurs in most disciplines find this boundary impassable due to the amount of technical expertise required, the inaccessibility of the discipline itself, or the barriers to publication imposed by the professional forum. Ornithology, however, has provided many intermediary forums (for instance, *American Birds* and *Audubon Field Notes*, both semi-professional publications, and state ornithology journals) through which amateurs can overstep the bounds and enter into a "foreign" discourse (Porter, 1986). It is within these limited forums that amateurs are given the privilege to speak as professionals.

What Forces Bind Discourse Partners Together? Though professionals participate in multiple organizations and must satisfy reviewers, editors, and established scientific paradigms, the social groups to which amateurs belong have fewer structures binding cohorts together. Amateurs may derive some sense of community through occasional national meetings and amateur publications, but their contact with fellow amateurs is restricted primarily to casual contacts or meetings with local bird clubs and nature societies. At other times, amateurs return to their own work and family worlds. Both amateurs and professionals have a sense of oneness with another "entity" (that is, the organization to which they belong), but the pull of that organization is fragmented in the case of an amateur. There is no obligation to pursue study, no sense of permanent commitment, no professional retribution if rules are violated or attachment temporarily or permanently severed.

What Is Their Worldview? From a personal, rather than a professional, perspective, amateurs may see themselves as part of the larger construct of "Nature," which implies an interrelationship among all living things. Thus, there is an identification with nature's creatures, including birds, which must confront the same basic obstacles as amateurs themselves, like the need to find food and shelter and the demands of reproduction. This worldview is the source of anthropomorphic interpretations of birds, one of the most prominent characteristics in amateur ornithological discourse and an example of ontological metaphors, which Lakoff and Johnson (1980) view as ways of making sense of world phenomena in human terms (p. 34). Ornithology in particular, through the study of bird migration and seasonal changes of bird populations, gives the observer a cyclic view of life as a continuing process of beginnings and endings, of life and death, of life's stages. Its study provides the observer with a microcosmic/macrocosmic worldview, much the same as for a gardener who may view plants' life cycles as indicative of her or his own, thus the verified therapeutic value of gardening (see, for instance, DeCrosta, 1981). Such worldviews can also incorporate the transcendent, in which birds become intermediaries, enabling the observer to contemplate supernatural truths. Amateurs also tend to have strong environmental concerns and focus on birds' well-being or lack thereof as symbols of the earth's own physical health. Roger Tory Peterson, unquestionably the most influential nonprofessional ornithologist of the century, noted that the route to environmentalism and conservation often originates in bird watching, since enthusiasm for birds often sparks larger environmental concerns and even careers in conservation or environmentalism (Fieth & Male, 1992).

There is also a segment of amateurs whose worldview involves more than simply an awareness of one's place in nature. Some see themselves as associated somehow with science, either as a former student of science or as a member of a scientific society. What drives them to "scientific observation" is a desire to "contribute something to science." Thus, they see themselves not as "amateurs," but as "amateur scientists" or "amateur ornithologists." In fact, the titles that amateurs assign to themselves are good indications of their worldviews. Roger Tory Peterson (1984) has identified a characteristic set of titles which signify one's place in amateur ornithology. In addition to ornithologist, one can be a bird watcher, birder, lister, ticker, twitcher, spotter, bander, or fancier. Since each of these titles has its own connotation in the world of ornithology, it behooves the speaker to call the listener by the correct name; for each of these names is indicative of a particular level in the pecking order of ornithological expertise.

What Is the Status of Their Knowledge and How Are Knowledge Claims Made? Since amateurs share in a very loosely bound discourse community without any uniformly recognized forums, the status of amateur knowledge varies depending upon the writer's persona. Since there are no educational prerequisites to be considered an amateur, amateurs' knowledge is derived primarily from direct observation and personal reading. There are no restrictions as to the kinds of knowledge they acquire; they observe and write about whatever is of interest to them. They may or may not have an organized methodology for their observations in the form of field notebooks, scanning techniques, shorthand symbols, procedures for counting birds, or audio or video recording. Knowledge claims are made by a mere statements of fact, perhaps supported with photograph, recording, or field notes. Discourse then tends to be descriptive, rather than argumentative; textual and pictorial, rather than graphical; narrative, rather than rhetorical.

How Do Amateurs Use the Same Knowledge as Professionals? Amateurs use ornithological knowledge to obtain personal satisfaction from and enjoyment of birds. They consider each observation and each bit of information acquired as another addition to their own accumulation of knowledge about the natural world. Some also use this knowledge to gain insight into the presence of a supreme being. Others quantify this knowledge in the form of checklists of birds, which amateurs accumulate from state to state, country to country, year to year throughout their lives. Thus, a birder's mark of expertise is the ability to acknowledge a "life list of 1,000 species," much like professionals' use of the honorific title "Doctor," though in the amateur's case the reward is a higher personal regard, rather than professional recognition. Others use their knowledge to form bird clubs and state ornithological societies, which function both as a social force to unite amateurs of similar interest and as a means to share the bits of knowledge each member has acquired. This sharing can occur both orally at club meetings and in written form as part of club newsletters and state ornithology publications, the primary discourse modes for amateurs. In particular, state publications are an apparent link to the professional, particularly when professionals edit and review submissions and the title "journal" is used, thereby giving the publication a professional aura. Still others use their knowledge of birds to advance the causes of conservation and environmentalism. For instance, amateur-compiled seasonal bird checklists and counts may relate changes in bird populations to a disappearance of forests or shoreline.

The Major Characteristics of Natural History Writers and Their Discourse: Two Case Studies

Case studies of two nineteenth-century natural history writers, Henry David Thoreau and John Burroughs, describe the features characteristic of natural history writers and their discourse. The following biographical sketches show the appropriateness of each of these writers as models for the discourse. Most importantly, Thoreau and Burroughs, like no other early natural history writers, were looked upon as guides and models by nineteenth-century aficionados of birds—and even cited by early American ornithologists. Thus, a study of the writings of Thoreau and Burroughs is an appropriate starting point for a study of American ornithologists' discourse. In addition, basing a model of natural history discourse on nineteenth-century writers is valuable because of the strong influence of nineteenth- and early twentieth-century natural history discourse on American ornithology, as will be shown in chapter 4. The resulting discourse model will better reflect natural history writing in American ornithology.

Henry David Thoreau

The activities of Henry David Thoreau as amateur scientist and naturalist (Harding, 1981) and his many references in his writings to the birds of New England (Cruickshank, 1964) make his writing a fitting choice for a natural history discourse model for ornithology. His *Maine Woods* included a bird list, and his natural history essays included observations of birds, as well as reflections about their place in the universe. His connection to ornithology is best seen in his *Notes on New England Birds* (1910) and his numerous references to birds in his journals, both of these works edited by Francis H. Allen, also an ornithologist. *The Auk* likewise acknowledged Thoreau, though authors were quick to point out Thoreau's literary and mystical inclinations. J. A. Allen, one of the AOU's founders, 28-year editor of *The Auk*, and major spokesperson during the AOU's first quarter-century, favorably reviewed Thoreau's *Notes on New England Birds*. As late as 1940, a "Recent Literature" entry in *The Auk* reviewed Henry Seidel Canby's biography of Thoreau, the issue's only non-ornithological review entry. Interestingly, the reviewer mentioned birds only once in passing, the remainder of the review discussing Thoreau's philosophical and transcendental beliefs (Wade, 1940).

Though Henry David Thoreau preferred calling himself a transcendentalist, commentators on Thoreau's activities as amateur scientist and naturalist have added many other titles, including meteorologist, botanist, geologist, entomologist, phenologist, ecologist, ethologist, limnologist, and conservationist. Also pervading Thoreau's 25 years of journal entries were his observations and commentaries on birds, commentaries that dated back to 1832 (Griscom, 1949, p. 4). These journals contained 8,433 entries on ornithology (Harding, 1981, p. 48), adequately justifying the added title of amateur ornithologist to Thoreau's list of accomplishments.

Thoreau's family was apparently the catalyst for his ornithological interests. Harding (1982) noted that as a child Thoreau became interested in bird songs through his mother's encouragement. He had also read about birds in notebooks of his brother John, whose interest in ornithology predated Henry's (pp. 10–11, 19). Once Thoreau's interest was sparked, he began filling his journals with notes on bird behavior, field marks, migration, nesting data, seasonal and extralimital occurrence, foraging, and bird song (Cruickshank, 1964). He acquired birds' skins and collected birds' eggs and nests, as was the custom of his day. Evidence of his attempt to systematize his records remains in an appendage to *Maine Woods* (Thoreau, 1906b, Vol. 3, pp. 347–349), where he listed the birds identified during his visit there.

Early on, Thoreau transferred these observations to his essays. "Natural History of Massachusetts" (1842), "A Walk to Wachusett" (1843), and "A Winter Walk" (1843) all include references to birds, but these references are more reflective and emotional than objective observation. His transcendental inclinations paramount, he rather sought spiritual laws from the birds he observed. Thoreau (1906a) readily admitted this in his journal by characterizing himself as a "mystic, a transcendentalist, and a natural philosopher to boot" (vol. 11, p. 4), a comment occasioned by his declining an invitation to join the prestigious Association for the Advancement of Science. His reservations about science are further seen in that he "deliberately did without scientific instruments, frequently because he believed they gave a distorted picture of nature" (Harding & Meyer, 1980, p. 128).

Despite his transcendental and literary inclinations, a perusal of Thoreau's journals shows his later years more obsessed with empirical facts. Most significantly, his major scientific contribution "The Succession of Forest Trees," read before the Middlesex Agricultural Society in 1860, became the standard description of the process. Though this work on forest regeneration clearly attests to his accuracy and skill in the scientific method, later readers of his works have called into question Thoreau's ornithological abilities. Naturalist John Burroughs (1889) remarked,

> To the last, his ornithology was not quite sure, not quite trustworthy. In his published journal he sometimes names the wrong bird; and what short work a naturalist would have made of his night-warbler, which Emerson reports Thoreau had been twelve years trying to identify! . . . From his journal it would seem that he was a long time puzzled to distinguish the fox-colored sparrow from the tree or Canadian sparrow,—a very easy task to one who has an eye for the birds. (p. 35)

Adding to Burroughs' comments, F. H. Eckstorm (1908) complained that Thoreau misidentified nine of the birds on his Maine list. A *Nation* book review of Thoreau's *Notes on New England Birds* criticized the omission of common Concord species, inaccurate field marks and misidentifications (Science, 1910, p.129). W. L. McAtee (1939) of the U. S. Department of Agriculture biological survey considered Thoreau's observations on the adaptation of birds to their environment to be speculation and emotionalism. Burroughs (1922) complained further that Thoreau could not distinguish between the songs of hermit and wood thrushes, did not know the yellow-bellied sapsucker's hole-drilling habit, and naively thought that bitterns produced their pumping sound by belching up water (pp. 135–137). Finally, Thoreau's *Journal* editor Bradford Torrey (1906) recounted that Thoreau had stood for quite a while within a few feet of a bird, never knowing whether he saw a woodcock or a snipe. Commenting on Thoreau's unfamiliarity with such common and distinctive Concord birds as rose-breasted grosbeaks, downy woodpeckers, and indigo buntings, Torrey concluded somewhat sorrowfully,

> The truth appears to be that even of the commoner sort of birds that breed in eastern Massachusetts or migrate through it, Thoreau—during the greater part of his life, at least—knew by sight and name only a small proportion (p. xlvii). [Thoreau] even in his later [journal] entries leaves the present-day reader wondering how so eager a scholar could have spent so many years in learning so comparatively little. (p. xlv)

Francis Allen (1910) and Mary Sherwood (1962) have adequately responded to some of this criticism; however, given Thoreau's expertise, for instance, in studying seed dispersal and measuring accurately the depth of Walden Pond, his ornithological inaccuracies need further consideration. Placing Thoreau in the context of a young, still evolving nineteenth-century American ornithology, as described in chapter 2, softens the unduly harsh criticism Thoreau has suffered as a result of the publication of his observations of birds.

John Burroughs

Another prominent figure representative of amateur discourse is John Burroughs, whose popularity as a natural historian during the late nineteenth and early twentieth centuries is attested to by his 1.3 million bird books sold by 1933, ranking them among the most popular of all writings on ornithology (Mayr, 1951/1975). It was, in fact, Burroughs who, along with John Muir, established the nature essay as a popular genre of American literature (Foerster, 1923, p. 264). His upbringing on a New York farm during the first 17 years of his life helped to established this lifelong interest in nature, bolstered by work on his parents' farm, including strawberrying, apple picking, and maple sugar-making. After receiving a limited education, he taught during his early adulthood, and in 1860 began his lifelong work as a nature writer, producing a series of essays for the *New York Leader*. Though as a young adult he moved away from his parents' farm in order to accept a teaching position, he would return there to work every summer. Biographer Jack Kligerman (1988) cited 1863 as a turning point in Burroughs' life when the naturalist found a copy of Audubon's *Birds of America*, which "reoriented Burroughs's imagination and gave him a vocation" because Audubon placed the bird in an environment and its viewer in close proximity (pp. 12–13). It was at this time that Burroughs began his study of ornithology (Renehan, 1992, pp. 64–65). But not only ornithology fascinated Burroughs. He became an authority on the wildflowers of his native land, as well as delving into other forms of wildlife and the geology from which they evolved. At age 34 he had written about and studied nature sufficiently to produce *Wake-Robin* (1871/1895b), his first book on nature.

In 1873, after abandoning teaching and a government job in Washington, D.C., where he continued to pursue nature study in his spare time, Burroughs returned to rural New York, purchasing a 9-acre farm not far from his parents' home. There, in the Catskill Mountains, he spent the remainder of his life in close contact with nature, eventually devoting his full time to a profitable vineyard. This contact, however, was not of the scientist, but of the lover of nature, as Burroughs (1886–1914/1968) described it: "Studied the birds? No, I have played with them, camped with them, gone berrying with them, summered and wintered with them, and my knowledge of them has filtered into my mind almost unconsciously" (Vol. 17, p. 13). However, like the scientist, the natural history writer does not alter the facts of nature but, unlike the scientist, adds an emotive element to them: "To interpret nature is not to improve upon her . . . it is to have an emotional intercourse with her, and reproduce her tinged with the colors of the spirit" (Burroughs, 1871/1895b, p. xiv). An advocate of raw nature, with which Burroughs lived intimately both in his early life on his parents' farm and in later adulthood in his two homes in the Catskills, he would always despise the "nature fakers." These, he scoffed, were both professional specimen collectors, whose egg shells and bird skins could in no sense rekindle nature in the possessor, and those nature writers whose mingling of the real with the imaginary turned nature into falsehood.

As his writings grew, so did his reputation. By the early twentieth century, not only famous personages but also the general reading public flocked to his home in the Catskills, where Burroughs always cheerfully greeted even his many unannounced visitors. His insistence on personally answering all correspondents, as well as a veiled invitation to visit the naturalist in a Burroughs-approved biography (Renehan, 1992, pp. 277–278), contributed to the swarm of admirers. In his later life, he successfully worked the lecture circuit and was honored with awards for his nature writing. His publisher Houghton Mifflin made him the center of a nature cult through the initiation of the John Burroughs Society in grammar and high schools as a combination natural history society and book club (Renehan, pp. 256–257).

Though Burroughs was the first to document the nesting of several species of birds in New York (Kligerman, 1988, p. 20), he was not directly connected with either the rise of the American Ornithologists' Union or with "technical ornithology," a term ornithologists of Burroughs' time used to distinguish themselves from amateurs. Yet the amateur naturalist was respected by the scientist-ornithologists of his time; for he demonstrated a scientist's approach to his observations of nature, evident in his admonition to his readers

> to see no more and no less than is actually before you; to be able to detach yourself and see the thing as it actually is, uncolored and unmodified by your own sentiments or prepossessions. In short, to see with your reason as well as your perceptions, that is to be an observer and to read the book of nature aright. (Burroughs, 1886–1914/1968, vol. 14, p. 238)

That his works were favorably reviewed in *The Auk* is partial acknowledgment of this respect. Unlike Thoreau, Burroughs was considered an ornithologist by ornithologists themselves, most notably by AOU spokesman J. A. Allen in 1903. Even during an attack on Burroughs in an 1886 reply to the naturalist's criticism of ornithologists' egg and specimen collecting in a *Century* magazine article entitled "Bird Enemies," J.A. Allen (1886) acknowledged that Burroughs "has even come to be looked upon as somewhat of an ornithologist, not only by the general public, but by ornithologists themselves" (p. 142). Ornithologists in effect symbolically raised Burroughs to the status of ornithologist by including his obituary in the April 1921 *Auk*. Witmer Stone (1921), the journal's editor, concluded in the obituary that Burroughs had "held a prominent place in the development of American ornithology," declaring the impossibility of estimating the number of people drawn into ornithology through his influence (pp. 317–318). Jean-Francois Lyotard (1984) believed that influence like Burroughs' is an essential characteristic of the scientist. For its own reproducibility, science needs student-adherents who in turn become disseminators of scientific knowledge; otherwise, the ongoing argumentation needed to perpetuate science would be stifled (p. 24).

Linguistic features of natural history discourse

As shown above, both Thoreau and Burroughs are appropriate subjects not only because of their experience as natural history writers, but also for their connection with American ornithology. The four selections chosen for this linguistic analysis of natural history discourse, two each by Thoreau and Burroughs, are likewise appropriate selections for this analysis of American ornithologists' discourse. After a brief description of each of these selections, the results of their linguistic analysis will be discussed.

In the first selection by Thoreau, "Natural History of Massachusetts" (1842/1980a), the naturalist reflected on a series of reports on the plants and animals of Massachusetts published for the state legislature and offered his own personal observations of nature. After asserting the value of nature study to the human spirit, Thoreau spoke briefly about one of his favorite topics, the cricket, before beginning his discussion of Massachusetts birds. His narrative described the birds present during each of the four seasons and included more lengthy descriptions of the osprey, bittern, and loon. Moving on to Massachusetts quadrupeds, he focused on the behaviors of the muskrat and fox. Then, after describing the kinds of fish one might expect to catch in the state, Thoreau briefly discussed snakes and turtles. Finally, a short discussion of flowers and trees led quickly into another favorite topic, the crystalline beauty of vegetation in winter. The 1,000-word section chosen for the linguistic analysis was, of course, his discussion of the birds of Massachusetts.

Thoreau's "The Succession of Forest Trees" (1860/1980b) is a particularly appropriate selection for analysis since it is the closest the naturalist came to writing a piece of scientific discourse and is still considered the standard description of seed dispersal. Because it is natural history discourse written for a scientific purpose, it parallels quite well the kind of discourse appearing in the early volumes of *The Auk*. The essay was first delivered as an address to the Middlesex Agricultural Society of Concord, Massachusetts, in answer to the question of how an oak forest springs up in a freshly cut pine tree lot, and vice versa. After explaining how pine seeds are dispersed by the wind into neighboring wood lots, Thoreau described the dispersal of heavy seeds and nuts, like oak acorns, into the pine forest by birds and squirrels. He interspersed his explanations with his own observations and experiments to show that the dispersal was accomplished yearly by wind and animals, rather than by the commonly held belief that the new saplings germinated from seeds long dormant underground.

The first of John Burroughs' selections, "Birds'-Nests" (1871/1895a), which originally appeared in *The Atlantic Monthly* (Burroughs, 1869), also has connections to scientific discourse because it offered vivid, objective descriptions of the nesting behavior of birds. The large number of nests observed by Burroughs from such a wide range of species from song birds to hummingbirds to eagles attested to his skill as a naturalist. His descriptions of the nesting activities of breeding birds showed the thoroughness and detail of a professional ornithologist. The topics covered included the preferred habitats and locations for nests, building of nests and nest cavities, types of materials used in nest building, manner in which birds feed their young and keep their nests clean, duties of both sexes at the nest, effects of the coloration of birds on nesting success, secrecy of nest-building, mating displays, and the effect on nesting of the death of a mate. Burroughs reprinted the article in his first nature book, *Wake-Robin*, whose value to early ornithology is apparent by the glowing review given it in the 1877 Nuttall *Bulletin*, which would soon become *The Auk*, the journal of the American Ornithologists' Union (Burroughs's, 1877, p. 48).

The reaction by ornithologists to Burroughs' second selection, "Bird Enemies" (1885), which first appeared in the *Century* magazine, showed that the nature writer's opinion held sway within ornithological circles. The article, which added specimen collectors to the list of birds' enemies, produced a quick retort from the *Auk* editor, J. A. Allen (1886). Burroughs intended to show the dangerous lives birds lead by recounting some of his personal observations of birds in dangerous situations. Cats, mice, snakes, and even other birds destroy birds' nests, eggs, and young. Birds' fragile bodies can become entangled in items such as nesting materials and caterpillar silk. Birds are susceptible to diseases and parasites and are hostage to the elements. But among birds' worst enemies, said Burroughs, are the specimen and egg collectors, whom he called "human weasels" (273), whose nest-robbing in the name of science should cease.

The linguistic analysis of these selections was made according to the methodology described earlier in this chapter. Table 3.1 shows the results of the analysis; occurrences listed for each feature are averages of the occurrences for the four natural history selections analyzed. The analysis suggests that the following linguistic features are characteristic of natural history discourse:

1. Average sentence length is moderate. (See Williams [1989, p. 111] for guidelines.)
2. Metadiscourse referencing author's thinking and writing structure is used.
3. Personal pronouns are used frequently, particularly the first person singular *I*.
4. Technical terminology is absent, and nominalizations are used only minimally.
5. Though the writer's persona is evident by the presence of personal pronouns, persona is occasionally obscured by the use of passive voice and place holders like *it*, *one*, and *there*.
6. Forms of the verb *to be* are often chosen (in the selections, 34% of the time).
7. Complex clauses are used frequently. (In the selections, 57% of the sentences contained complex clauses.)
8. Relative clauses are the most common form of complex clauses; minimal use of causal clauses reflects the absence of scientific argumentation.
9. A small percentage of subjects and verbs are interrupted by intervening words.
10. For making factual claims, natural history discourse uses a wide variety of evidentials, though citations are noticeably absent.

Format and Structure of Natural History Discourse

Natural history writers are primarily essayists, whose works appear generally in popular magazines, newspaper articles, and full-length books. In modern-day works, some lengthy descriptions may give way to visuals, but the selections by Thoreau and Burroughs were written in continuous text without headings or graphics. As professional writers, they relied on words to convey nature to their readers. However, when Burroughs reprinted his article "Birds'-Nests" in *Wake-Robin* (1871), he included one reproduction of a drawing by Louis Agassiz Fuertes of a perched Osprey. The bird is incidental to the article, which devotes only a paragraph to the raptor. Similar pictorials are dispersed throughout the book. In general, graphics in natural history discourse will function similarly; that is, they are commonly

TABLE 3.1.
Linguistic Analysis: Natural History Discourse

Linguistic feature	Avg. occurrences
Average sentence length	29.4
Metadiscourse	
refer to own thinking	2
refer to structure of writing	1
refer to reader's act of reading	0
Nominalizations	5
Technical terms	0
polynomial technical terms	0
Personal pronouns	20
Use of definitions	<1
Sentence complexity	
subjects as place-holders	6
passive voice	12
to be	20
reporting verbs	3
active verbs	58
% simple sentences	31
Complex clauses	
relative noun	14
relative adjectival	9
time	8
causal	2
place	<1
conditional	6
Subject-verb interruptions	
% interruptions	15
number of interrupting words	59
Evidentials	
reliability of evidence	7
belief	1
induction	1
hearsay	1
citation	0
deduction	6
hedging	3
sensory evidence	4
expectations	6

drawings, photographs, or paintings of birds in unusually bright colors or interesting poses, often merely to complement the text or to elicit emotional appeal.

EXPERIMENTAL SCIENCE DISCOURSE MODEL

Compared to the natural history discourse model, the experimental science model is characterized by a greater reliance on authority and is driven more significantly by external forces, as shown by the following premises and linguistic and structural features.

Rhetorical Considerations

Here again, answers to the five questions below will clarify the context within which experimental science discourse occurs[4].

Who Am I and For Whom Am I Speaking? The persona of the professional has an aura of authority, acquired through accumulated education, experience, past publications, title, job description, and workplace prestige. For a scientist, acquisition of such an aura has come after many years of study and with much struggle through extensive educational training, the rigors of experimental research, and the rocky road to publication. As scientist John Ziman (1981) commented: "The student who gets through a research degree is a true specialist, having learnt a hundred times more about a thousand times less than any ordinary person would want to know" (p. 125). These struggles and the resulting rewards of publication, recognition, and position give the professional the privilege to speak to and on behalf of the scientific community. There are many forums provided for such communication, including numerous general and specialized professional journals, conference presentations and proceedings, and even popular forums such as *Scientific American* and *National Geographic*. Thus, implicit in a scientist's self-identification is a realization of privilege, priority, authority, and even insularity and exclusivity, which may arise when self-trained outsiders attempt to enter into scientific discourse and the knowledge-making process.

In one instance, a university ornithologist was recently involved in a disagreement over the location of the state's ornithological records and collections. The ornithologist, whose department was the present depository, opposed the proposal of officials who wanted the state's science museum designated as the official state depository. Officials argued that the museum was more centrally located than the university and was thus more accessible to the state as a whole. The principal argument of the ornithologist focused on the fact that museum workers were not "professionally trained" for such responsibility; that is, they were not advance-degreed zoologists. Due to the ornithologist's opposition, the change was not made.

In light of this self-image, professionals see their audiences as peers whose educational background and expertise allow them to

- understand scientific writing and the field's research goals
- recognize experimental models accepted by the scientific community
- have a sense of past research methodology and theories
- understand technical terminology
- function as insightful readers and peer reviewers
- appreciate and intelligently evaluate new, perhaps innovative or controversial methods, techniques, theories, and models, the acceptance of which allows the proposing scientist either an initial place or a higher position in the scientific community

Thus, power and authority play important roles in a professional's perception of audience (Latour, 1987). In essence, professionals attempt to place themselves intellectually above their audiences so as to argue from a position of authority (Bartholomae, 1985). But it is also an audience from whom professionals must receive acceptance through citation; otherwise, their discourse, which announces their proposed methods, theories, and models, is buried in the accumulation of journal writing, never to be referenced afterward (see Zuckerman & Merton, 1973).

What Forces Bind Members Together? Unlike amateurs, professionals are part of a superstructure which bestows on them titles such as ornithologist or ichthyologist. This superstructure includes professional meetings and conferences, publications such as journals and proceedings targeted to a specialized audience, and a limited realm of prime "physical space" (that is, employment within academia, industry, and government) in which the professional functions.

Such affiliations with academic departments and professional work groups also strengthen ornithologists' bonds with the community by providing the opportunity to work with colleagues having similar interests. The act of publication, and more importantly, the process of subsequent citation by one's colleagues, provide both the psychological and emotional reward of having one's work recognized by one's peers. In business and industry, a successful outcome from a research and development project may lead to corporate advancement.

Professionals usually have strong commitments to the professional organization, that entity with whom they are identified as "ornithologists." Bonds can be strengthened or weakened depending on the type of degree and the institution that empowers the use of exclusive titles. For example, an added title of "Doctor" or "Ph.D." makes more cohesive the bond among professionals, since "naming" confers a sense of attachment to the social group, much stronger than the amateur's sense of social self.

What Is Their Worldview? John Ziman (1981) summarized the motivation to be a scientist as follows:

> The aspiring young scientist dreams of "pushing back the frontiers of knowledge." There is a mysterious desire to "explore Nature" or to "satisfy human curiosity." Pressed to give a more precise motive, the aspirant may murmur something about "making two blades of grass grow where one grew before." For some there may be an unacknowledgeable ambition to win fame and fortune with a Nobel Prize. For others it may be just a wonderful opportunity to practise some absorbing craft, like mathematical manipulation, electronic design, or messing about with bubbly coloured liquids and glassware. Some take to science because there [*sic*] are jolly good at it and have no confidence that they could succeed anywhere but in academia. (pp. 123–124)

Thus, professionals' worldviews are interwoven with their position, reputation, and advancement, as well as with their own research goals. Professionals also see themselves as "contributing to science"; but their contribution is not an individual one, like the amateur's. It is rather part of a larger scheme of scientific inquiry enveloped in an interconnected web of scientific thought, which, if incorporated into scientific knowledge by referencing and replication, perpetuates the professional's scientific persona.

Several philosophers of science have proposed theories that clarify professional worldviews. Karl Popper (1959) viewed science as a rush to the laboratory in an attempt to refute a proposed theory. If not contradicted, the theory becomes part of scientific knowledge. Taking a social view of science, Thomas Kuhn (1970) suggested that a scientist proposing new theories must compete with fellow professionals as resistance to novel theories is encountered. Such controversy, if persistent and irresolvable by current paradigms, ultimately causes a reevaluation, or a state of uncertainty, before a new paradigm becomes established returning science to normal. Irme Lakatos (1978) believed that science operates through research programs. Those programs, which are successful in perpetuating topics for research, will continue to attract scientists and thus will be the centers of scientific advancement, until a program with new, better, and more easily fundable research goals supersedes the older program. Bruno Latour (1987) distinguished between cold science, in which established theories are passed on as if in sealed "black boxes" holding what is acceptable scientific knowledge, and warm science, in which these boxes are opened in order to alter theory, placing science in a state of flux. Finally, Marcello Pera (1994), arguing that science requires an imprecise methodology, proposed a dialectical model for scientific argumentation, which includes scientific rhetoric as a central component. Through dialectic, the scientific community represented by rival groups eventually arrives at a consensus regarding the fate of potential new theories.

Professionals then are part of a community of associates who, though they may disagree on methodology or theoretical issues, have a unified worldview based on agreement concerning past scientific knowledge, basic methodology as outlined by experimental design principles, and statistical means of argumentation. They must not just maintain their place, but advance in their profession because scientific research is their livelihood, providing both income and self-identity, whereas amateurs have their own largely non-scientific careers, which pull them away from the ornithological world.

However, a mutual attachment to birds forges a unique relationship between professional ornithologist and amateur. There are ornithologists who write bird-finding guides, participate in bird clubs and Audubon societies, join in bird counts, conduct field trips for amateurs, and even run birding telephone hot lines. Interestingly, ornithologists, like amateurs, can often name a particular bird that sparked their initial interest in ornithology, much as amateurs do (see Kastner, 1986). Yet unlike amateurs, professionals see birds as subjects to be manipulated, measured, and experimented upon; thus, one of the major and most evident rifts between amateur and professional ornithologists through the years has been the collection of specimens for laboratory study. A reading of the methodological sections of *The Auk* would enrage many an amateur: caged juncos exposed to temperatures up to the point of freezing in order to study bird metabolism; live hawks stuffed into glass jars for determining reactions to various mixtures of air; turkey vultures placed in boxes allowing for dangling legs, which are then coated with varied mixtures for studying respiration—all this accompanied by blunt language like "the subjects were killed and dissected." As a result, the impartiality of professionals toward nature as objects of study and the need for scientific objectivity ultimately override personal or anthropomorphic interests in birds, perhaps lurking somewhere in the back of the researcher's mind.

What Is the Status of Their Knowledge and How Are Knowledge Claims Made? Professionals share in numerous discourse communities and sub-communities, each with its own recognized forums, which have specific mechanisms awarding participants the right to speak to the community. Mechanisms such as peer review add a competitive element to scientific endeavor, one in which professionals must not only accept criticism, but also react to and accommodate themselves to their colleagues' concerns and reservations. Thus, published discourse acquires status in its own right. For the professionals themselves, the status of their knowledge is influenced by education, position, and publication; that is, the more prestigious the education, the higher the position; and the greater volume of publication, the higher status given to their knowledge claims. Though amateur study has no restrictions, the knowledge that professionals generate must be located within recognized research programs and in conjunction with opportunities for funding. Through higher education, professionals acquire distinct methodologies and theoretical approaches, which provide additional support to knowledge claims. However, because methods and theories may be confronted by counter methods and theories, knowledge claims must be supported by argument and, perhaps, ultimately by rhetoric. Discourse then tends to be expository and descriptive; graphical, highly structured, and skimmable; argumentative and rhetorical.

How Do Professionals Use the Same Knowledge as Amateurs? Like amateurs, professionals make assertions concerning ornithological knowledge, but on a much more sophisticated level and with more wide-ranging implications. For instance, empirical observations may indicate changes in bird populations, which in turn reflect the status of a species, the health of the environment, or an evolutionary event in process. Others use data as part of computer and statistical models to reject or accept theoretical hypotheses, thus helping to shape new black boxes and articulate new scientific paradigms. As a result of publication, knowledge also becomes professionals' means for both advancement in their profession and peer recognition for their contribution to scientific knowledge.

The Major Characteristics of Experimental Science Writers and Their Discourse: Two Case Studies

Logically, examples of experimental science discourse that will contrast best with natural history discourse are those that deal with the most technologically complicated and technically restrictive research in American ornithology (that is, examples at the opposite end of the spectrum from natural history discourse). For this, we again return to systematics, from which experimental science in ornithology has evolved over the past four centuries. Traditionally, avian anatomy has been used to reconstruct the family tree of birds, but genetics has now begun to play a crucial role. Just as human genetic research is both medically promising and societally controversial, avian genetics research during the last decade has not only held the promise of understanding better the interrelationships among bird families, but also rekindled centuries' old disputes over genus-species relationships. In the process of developing reliable methodologies for this genetics research, contemporary systematists have used a number of biochemical techniques, including amino acid sequencing, electrophoresis, mtDNA sequencing, and DNA-DNA hybridization. The last two techniques in this list, mtDNA sequencing and DNA-DNA hybridization, are the methodologies used in the experimental science discourse analyzed in this study.

In an introduction to a volume on contemporary research in American ornithology, ornithologist Ernst Mayr (1983) commented: "It can hardly be questioned that no other recent advance in biology has had a deeper and broader impact than molecular biology" (p. 15). For ornithology, DNA research reevaluating the taxonomic relationships among birds has had a great impact upon ornithological knowledge. Mayr (1983) has also described the controversial nature of this branch of ornithological research (p. 8). Myers (1990), as well, has documented the complex peer review process in the publication of a controversial genetics-based article on evolutionary biology, which was accepted only after the writer lowered the status of his claim (pp. 63–100).

Such genetics research represents American ornithology farthest removed from the amateur, aimed rather at a narrow group of professionally trained readers. The use of sophisticated equipment, such as electron microscopes and DNAlyzers, and the creation of technical terminology, like *infraorder* and *parvclass*, terms new even to long-time professionals, make DNA research an appropriate topic for this analysis of experimental science discourse in American ornithology. The resulting discourse is particularly appropriate for developing the experimental science model since it represents ornithological discourse in its most specialized and exclusive form, as close as one might come to "pure" experimental science discourse produced by American ornithology. That is, true models of natural history and experimental science discourse will be exclusive of one another. As Myers (1990) has commented, a characteristic of professional discourse and nonscience discourse is that one is not entered into by the other (p. 145).

In addition, the topic of DNA research is particularly appropriate for this study, since it represents the continuation of the study of systematics shown by chapter 2's historical overview as the branch of ornithological study that gave rise to the professional ornithologist. That is, it is a contemporary manifestation of taxonomic research, but one that can now be conducted only with sophisticated equipment, extensive professional training, complex analytical ability, and an array of rhetorical skills. Finally, the evolution from simple specimen study for cataloging birds to complicated DNA research for ordering birds into a uniform framework parallels the evolution of methodologies for identifying natural phenomena. That is, the Linnaean classification system based on easily observable physical features eventually gave way to complex natural classification systems, as, for instance, the change from classifying minerals physically to classifying them chemical-analytically (see Daniels, 1967, p. 153).

The systematic ornithologists

Selections by two contemporary systematic ornithologists, named here as ornithologists A and B, demonstrate the linguistic features of experimental science discourse.

At the time of publication, ornithologist A, the sole author of two selections, was a member of a biology department in a major university and was also associated with a major museum. Ornithologist B was associated with another major museum, which included a bird section among its collections. B was the sole author for one of the selections and was the principal author of the second selection, whose co-author was associated with the same museum as B. Both authors received their Ph.D. degrees during the mid-1980s from major research universities and have co-authored papers with prominent systematic ornithologists. Ornithologist A has been publishing genetics research since 1987, B since 1984. Since then, both authors have single- or co-authored 17 genetics articles, most of them in major journals in the biological sciences. Thus, both scientists bring to their discourse prior knowledge of the kind of writing needed for scientific publication and the stylistic requirements of professional journals in the biological sciences. Each of the four articles analyzed here was published within the last 10 years in a major professional journal of ornithology.

Linguistic features of experimental science discourse

Four empirical studies were chosen to describe the linguistic features of the experimental science discourse model. Three of these studies, two by ornithologist A and one by ornithologist B, described the use of DNA-DNA hybridization to reclassify birds taxonomically according to their genealogical relationships. The second study by B used mtDNA sequencing methodology.

So that the first three selections analyzed will be understood more fully, here follows a brief, though admittedly reductive, explanation of DNA-DNA hybridization.[5] In this hypothetical example, the researcher is attempting to determine the genealogical relationship between two species of birds. DNA is first extracted from red blood cells of a specimen of each of the two species under study. Heating the DNAs of both species to the melting point causes the two chains of each species' DNA to separate. Next, by placing one DNA chain from one species and a melted chain from the other species together in the same mixture, which is then cooled, the two chains will try to realign with one other. The new hybrid DNA is then reheated a few degrees at a time in order to split it apart again. After each increase in temperature, the DNA fragments that have melted at that particular temperature are collected and measured. By comparing the amount of the hybrid DNA melting at each temperature, the amount of genetic difference between the two species and ultimately the length of time since the two species split from each other can be determined.

The first selection by ornithologist A reported the results of DNA-DNA hybridization techniques on 13 species which represented all of the major groups of songbirds found in the Western Hemisphere. The family tree derived from A's DNA study consisted of four branches: tanager, sparrow, warbler/oriole, and finch. As expected, the proposed taxonomy suggested a number of new relationships. The results of the study also suggested that the use of feeding behavior as one means of distinguishing relationships among birds may be less reliable than previously thought.

A's second selection described a hypothetical study which questioned the requirement that a constant average evolutionary rate be established before using DNA-DNA hybridization. Briefly, researchers have calculated the amount of time since the splitting of two lineages of birds by using a formula that multiplies (a) the differences in the amount of temperature change needed to melt hybrid DNA vs. normal DNA times (b) a specific number in millions of years based upon some geologic event. For this formula to work, it has been assumed that the average rates by which species' DNA evolves over time must be identical. Researchers have verified this constancy by conducting a test called the relative-rate test. The premise that genes evolve at a constant average rate has been a key advantage of DNA methodology since morphological and anatomical change, traditionally used to develop taxonomic relationships among birds, does not necessarily occur at a constant rate. In the selection analyzed here, ornithologist A argued that, if pair-wise statistical methods and heuristic algorithms are used, then DNA data may still be used despite varying evolutionary rates. In A's hypothetical study, though rates were chosen randomly, the data using A's methodology produced the correct branching patterns. The article concluded by discussing some of the problems inherent in the use of DNA methodologies.

The first selection by ornithologist B also dealt with the constant-rate premise, as well as with random events that may also play a role in genetic evolution. B first created a hypothetical evolutionary tree composed of three species of birds and then statistically analyzed several hypothetical sets of data using a computer model that evaluates a random mode of evolution. Variables in the model included the time over which random events occurred and the evolutionary rates of the hypothetical family of birds. A series of graphs illustrated the probabilities of accurately deriving correct branching patterns for the three hypothetical species given several sets of variables, the values of which were set by B. The study supported the value of biochemical analysis, but showed many circumstances in which genealogical information would not be recoverable. B concluded by warning against overinterpretation of biochemical data and by suggesting that multiple approaches, including more traditional morphological analyses, might be a more valuable methodology.

B was the principal author of the second selection, which listed one additional author. After some of the shortcomings of DNA-DNA hybridization were described, a somewhat different DNA methodology, mtDNA sequencing, was proposed to reevaluate the relationship among three groups of tropical birds, whose genealogy had already been described using DNA-DNA hybridization. Briefly, mitochondrial DNA (mtDNA) sequencing involves statistically analyzing the genetic information on segments of DNA. This information appears as a specific sequence of the four bases, called adenine (A), thymine (T), cytosine (C), and guanine (G). Mitochondrial genes, one type of avian gene, are carried on DNA in an organelle called a mitochondrion located in the cytoplasm of a cell. Because of their rate of change and maternal origin, these genes may be used to trace genealogy in the following way. Researchers are able to break DNA into meaningful fragments by using restriction enzymes, a series of proteins that can read DNA. After these fragments of DNA are obtained from each species under study, the various genetic sequences and the number of fragments obtained for each of these species are compared and genealogical trees proposed. B's study confirmed the genealogy already suggested by prior morphological and DNA-DNA hybridization studies. Though B's study replicated prior results, it was an important confirmation because the results were obtained by a less controversial DNA methodology.

Table 3.2 gives the results of the linguistic analysis of these articles. The major linguistic features of experimental science discourse, as suggested by the analysis, are as follows:

1. Like the natural history model, the average sentence length is moderate (See Williams, 1989, p. 111, for guidelines), though somewhat shorter on average than natural history discourse.
2. Metadiscourse is used, including reference to the reader's act of reading, which indicates the need to refer professional readers to related studies.

3. The extensive use of nominalizations; technical terms, including polynomials; and definitions reflects the specialized nature of experimental science.
4. Using passive voice (somewhat more extensively than natural history discourse) and place-holders as subjects, in conjunction with the very limited use of personal pronouns, effectively obscures the presence of the writer.
5. Despite an average sentence length similar to that of natural history discourse, there is a decrease in the percentage of complex clauses. (The 90% increase in the number of words used between subjects and verbs and the presence of polynomial technical terminology account for sentence lengths approaching those of natural history discourse.)
6. Like natural history discourse, relative clauses predominate, though, in general, clausal usage decreases in experimental science discourse.
7. The 87% increase over natural history discourse in the number of words used between subjects and verbs, represented in large part by adjectival prepositional phrases, reflects an increased need to define ideas and terminology.
8. The need to demonstrate scientific objectivity is evident by the predominant use of two evidentials, reliability of evidence and citations, and by the almost complete suppression of most other markers for making factual claims.

Format and Structure of Experimental Science Discourse

Writers of experimental science discourse have numerous genres available to them: reports, journal articles, reviews, conference proceedings, grant proposals. In stark contrast to natural history discourse, experimental science discourse is structured with headings required by the format specifications preferred by each field. Both writers A and B followed the modified IMRaD format required by their journals—Abstract, Methods, Results, Discussion, Acknowledgments, and Literature Cited. Ornithologist A also subheaded both Results and Discussion sections into meaningful subdivisions. Though A's second article does not include IMRaD formatting, this organization is implicit in the text.

More strikingly different from natural history discourse were the graphics used by both authors. These complex visuals indicate the specialization of the authors' topics and the sophisticated audience the articles address. With the advent of computerized graphics packages, the inclusion of such graphics displays has become a key component of scientific argumentation. Ornithologist A used three types of graphics in the first selection: one multiple line graph; three multi-column tables of statistical data, one of these a tightly packed, 26x14 single-spaced matrix of probability distributions; and

TABLE 3.2.
Linguistic Analysis: Experimental Science Discourse

Linguistic feature	Avg. occurrences
Average sentence length	27.9
Metadiscourse	
refer to own thinking	2
refer to structure of writing	2
refer to reader's act of reading	1
Nominalizations	19
Technical terms	35
polynomial technical terms	29
Personal pronouns	3
Use of definitions	14
Sentence complexity	
subjects as place-holders	4
passive voice	16
to be	16
reporting verbs	2
active verbs	35
% simple sentences	42
Complex clauses	
relative noun	8
relative adjectival	8
time	2
causal	3
place	0
conditional	5
Subject-verb interruptions	
% interruptions	28
number of interrupting words	112
Evidentials	
reliability of evidence	15
belief	<1
induction	1
hearsay	<1
citation	16
deduction	5
hedging	<1
sensory evidence	0
expectations	4

four avian genealogical trees. The second article contained two avian trees that included statistical data printed along each branch of the tree. In ornithologist B's first selection, there were six figures containing multiple 3-D surface-fill charts of probability distributions evaluating the accuracy of DNA interpretations, along with one scatterplot and one genealogical tree. The methods section was also interspersed with complex probability formulas. The second article by B contained four tables, including a full-page listing of statistical and DNA sequencing data.

A SCHEMA FOR ORNITHOLOGICAL DISCOURSE

Now, let us begin to construct a diagrammatic representation of American ornithological discourse models and the components that play a role in their interrelationship and change. Figure 3.2 illustrates natural history and experimental science discourse, the two poles of discourse modeled in this chapter. The line separating the two models represents the discourse variation responsible for distinguishable models. Implicit in the term "discourse variation" are not only the distinctive linguistic features presented in this chapter, but also both the social forces that give rise to varying plays for authority and power and writers' self-identity, which invents the varying audiences addressed. For pure models, this line is fixed. However, the length of the line will vary, given actual examples of discourse.

But are these the only viable discourse models for American ornithology? Chapter 4's historical overview of the topics and themes appearing in *The Auk* during its century of publication will suggest the need for a third model, which will necessitate a refinement of the schema proposed here. By what mechanisms did American ornithological discourse evolve to such a degree? How did it do so in such a relatively short time? To answer these questions, chapter 5 will take a closer look at *The Auk*, American ornithology's professional journal. Chapter 6 will reflect upon how American ornithological discourse evolved from natural history writing similar to that of Thoreau and Burroughs into a fully professionalized discourse, which can now include as a component specialized discourse like that analyzed in this chapter.

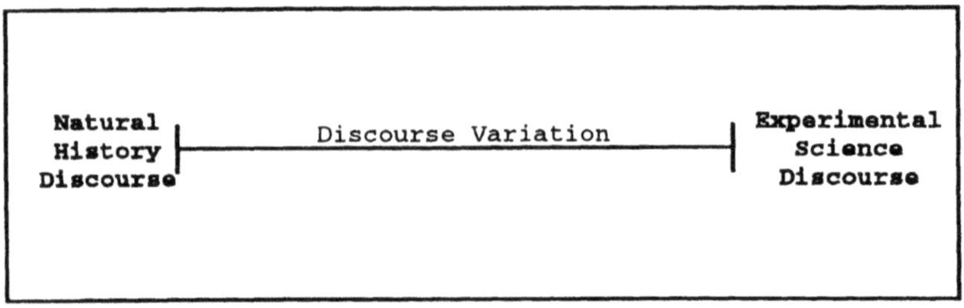

FIGURE 3.2. Discourse schema showing the two poles of ornithological discourse.

4

Diachronic Changes: The Discourse of Transition*

The natural history and experimental science discourse models described in chapter 3 are not isolated entities developing in a vacuum as opposite poles of discourse. On the contrary, there has been an interplay between these two forms of discourse as the science of American ornithology itself has advanced through time. The first half of this chapter presents evidence of this interplay of discourse through a historical overview of the topics and methodologies present in American ornithology over the past century. As will be shown, American ornithologists did not abruptly switch from natural history topics and methodologies to those of experimental science, but rather made a gradual changeover from one to the other as the science professionalized. Though this period of transition may not be defined by specific dates, since the writing associated with this transition evolved gradually though recursively, the period does have specific characteristics which can be described in a third discourse model. This chapter first documents American ornithology's gradual and somewhat protracted shift from natural history to experimental science topics and

* Part of this chapter has appeared previously in Battalio, J. T. (1996). The interplay between narrative, education, and exposition in an emerging science. *Journal of Technical Writing and Communication, 26,* 177–191. © 1996, Baywood Publishing Co., Inc.

methodologies and then offers a third discourse model—that of transitional discourse—to reflect this shift in theoretical and methodological perspective.

THE PERSISTENCE OF NATURAL HISTORY DISCOURSE IN AMERICAN ORNITHOLOGY

In order to clarify the transition from natural history to experimental science as American ornithology has professionalized over the past century, this chapter uses *The Auk*, the profession's major journal, as a historical barometer of this progression by surveying the topics and methodologies described by American ornithologists in a cross section of articles over the journal's history. The study categorizes the topics and methodologies described in all major articles[1] from all four issues of 11 volumes of *The Auk* beginning in 1890 and each 10-year interval thereafter (that is, 1900, 1910, etc.). The 444 articles represented in this analysis give a clear overview of the development of American ornithology and show the persistence of natural history themes and methodologies through much of the field's 100-year history, even as experimental science moves to the forefront. Given the two discourse models for natural history and experimental science described in chapter 3, we can by extension assert that the changes in topics and methodologies in American ornithology document corresponding changes in the discourse of American ornithologists over the same period of time. Applying this perspective to the history of the discourse itself, it appears that amateur discourse, as represented by natural history writing, has persisted as a legitimate form of writing by ornithologists at least through 1970, much longer than would be expected, given the continuing professionalization of American ornithology during the twentieth century.

In order to visualize the interchange between natural history and experimental science discourse, it would be instructive to consider the topics and methodologies found in each of these articles as part of a continuum from amateur to professional discourse. Such a continuum would provide a closer look at the evolution of ornithologists' writing during the past 100 years of American ornithology. One useful method is to place these 444 examples of discourse in the Continuum of Scientific Perspectives on Nature (Killingsworth & Palmer, 1992), described in the previous chapter (see Figure 3.1).

Table 4.1 shows the distribution of articles across the continuum[2]. Since all but 32 of the articles, or 93%, fall on either extreme, that is of natural history or theoretical science, an analysis of discourse in *The Auk* can focus on these 412 articles. Because they are distributed over the past century, the articles can be considered a representative cross section[3] of topics of interest to ornithologists over the years and reflect methodology in keeping with what was considered ornithological science at each respective period in time.

TABLE 4.1.
Ornithological Discourse in *The Auk*: 1890-1990 [number of articles (%)]

	Observe in field		Observe & measure in field		Manipulate in field		Manipulate in semi-natural conditions		Bring system into lab		Theor. sys. math & comp. models		Other		Total no. of articles
1890	17	(41%)	6	(14%)	0	(0%)	0	(0%)	17	(41%)	1	(2%)	1	(2%)	**42**
1900	14	(40%)	2	(6%)	0	(0%)	0	(0%)	16	(46%)	1	(3%)	2	(6%)	**35**
1910	24	(69%)	6	(17%)	0	(0%)	0	(0%)	2	(6%)	0	(0%)	3	(9%)	**35**
1920	20	(61%)	2	(6%)	0	(0%)	0	(0%)	8	(24%)	1	(3%)	2	(6%)	**33**
1930	20	(50%)	5	(13%)	0	(0%)	0	(0%)	7	(18%)	2	(5%)	6	(15%)	**40**
1940	6	(18%)	6	(18%)	3	(9%)	0	(0%)	14	(41%)	2	(6%)	3	(9%)	**34**
1950	6	(22%)	7	(26%)	0	(0%)	0	(0%)	7	(26%)	4	(15%)	3	(11%)	**27**
1960	4	(17%)	5	(22%)	1	(4%)	0	(0%)	10	(44%)	2	(9%)	1	(4%)	**23**
1970	2	(5%)	16	(38%)	3	(7%)	1	(2%)	16	(38%)	4	(10%)	0	(0%)	**42**
1980	1	(2%)	20	(29%)	2	(3%)	0	(0%)	17	(25%)	28	(41%)	0	(0%)	**68**
1990	1	(2%)	5	(8%)	1	(2%)	0	(0%)	3	(5%)	55	(85%)	0	(0%)	**65**
Total	**115**		**80**		**10**		**1**		**117**		**100**		**21**		**444**

Since the total number of articles varies from decade to decade and since the study is mainly concerned with the changes in types of discourse, this analysis will use percentages, rather than raw number of articles, per year. In addition, for the purposes of this analysis, the word "volume" will refer to one of the eleven volumes surveyed, rather than to any of the 109 volumes at random.

The following series of graphs clarify the relationship between the discourse of natural history and experimental science in American ornithology by showing that a definite swing in the pendulum away from natural history discourse has occurred only within the past decade. This change will be evident first by looking at the percentage of articles appearing at each end of the continuum (that is, natural history and theoretical science) and then by placing these two extremes side by side in order to view the interplay between the two types of methodology and, in essence, the two forms of discourse.

Ornithological Discourse at the Two Extremes of the Continuum

Figure 4.1 displays the natural history end of the continuum with two peaks clearly evident. First, the solid bars of the graph show the predominance of natural history observation during the first half-century of American orni-

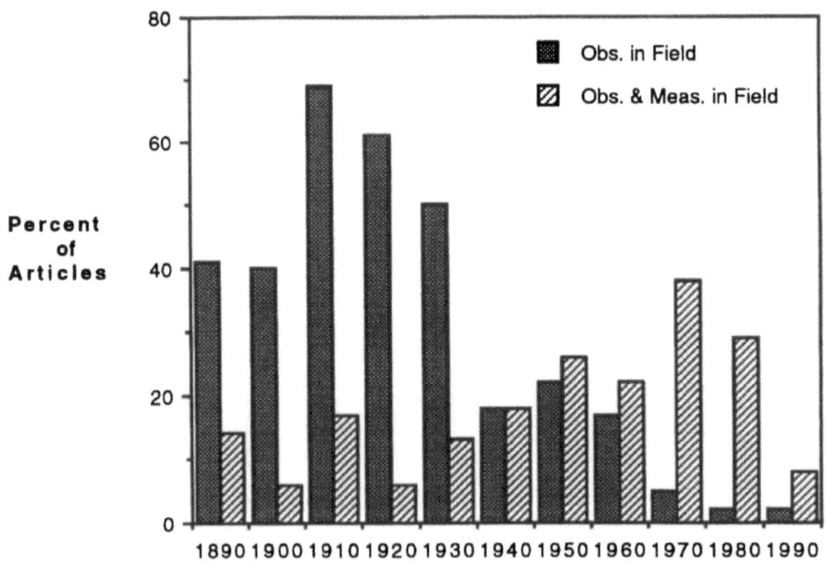

FIGURE 4.1. Interplay between field observation and measurement (N=195).

thology and particularly its dominance during the early part of the twentieth century. Most evident is the interplay between natural history observation and observation accompanied by measurement. Though direct observation predominates as the major methodology until about the 1940s, greater reliance on measurement in the field (striped bars) appears to make simple field observation obsolete by the last three decades surveyed.

Since measurement in the field must be accompanied by observation first, both of these functions are actually closely allied, both in methodology and presentation of discourse. Thus, for the purposes of this study of natural history versus experimental science discourse, it will be appropriate to group these two functions together under the heading Natural History Discourse, as shown below in Figure 4.2 (see the solid bars).

The significance of looking at ornithologists' writing from this perspective is that the percentage of natural history discourse, as shown by the black bars, remains at least average (mean = 47%) through 1970, with the exception of a slight decrease in the 1940 volume. That is, for most of the volumes 1890 through 1970, at least half of the discourse in any given volume is devoted to natural history observation and/or measurement. Significantly, natural history discourse does not begin to lose ground until the 1980 and 1990 volumes.

Given this overview of the progression of natural history methodology and discourse used by American ornithology, the next two graphics make similar analyses for the theoretical science end of the continuum. Figure 4.3 shows the progression of both laboratory work (solid bars) and theoretical

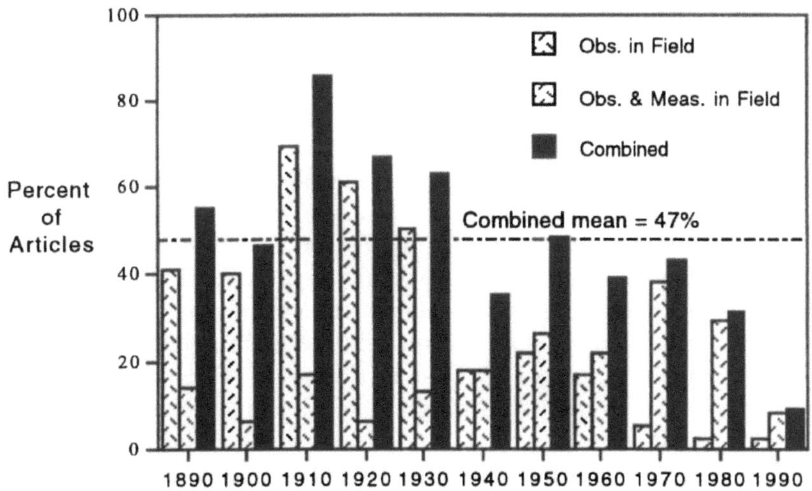

FIGURE 4.2. Natural history discourse (N=195).

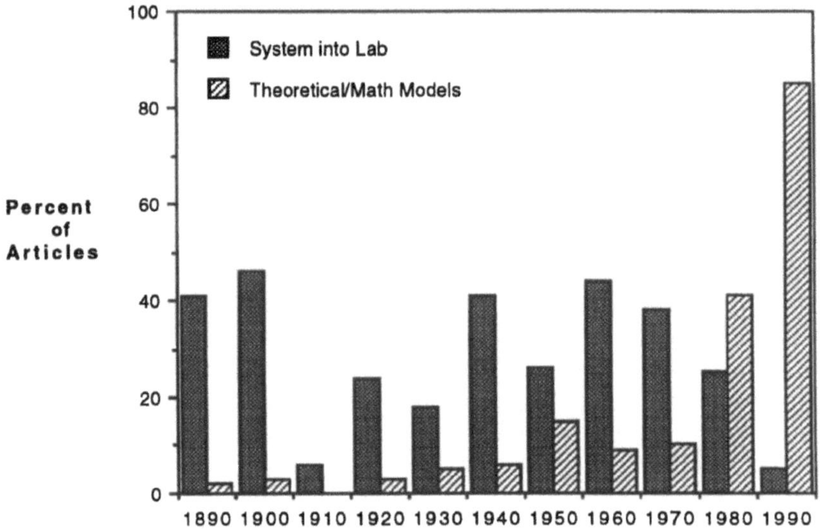

FIGURE 4.3. Interplay between the laboratory and theory (N=217).

discussions (striped bars) in ornithology. Laboratory study clearly dominates until after the 1970 volume. However, as mentioned in note 2, it should be remembered that the method used to classify articles was to place each article in the category farthest to the right (the theoretical science end of the continuum); that is, if an article, though based in the experimental laboratory, employs theoretical or mathematical models, it was categorized as a theoretical article. Actually, in 1980, 12 of the 28 theoretical articles (or 43%) were laboratory-based and, in 1990, 15 of 55 (or 27%)[4]. Consequently, ornithologists had not abandoned the laboratory for theoretical discourse to the extent shown in Figure 4.3. In addition, the decline in laboratory studies in the early part of the century is accounted for by the corresponding surge of interest in natural history, as shown in Figure 4.1 above.

Now, for the purposes of the study, Figure 4.4 combines percentages from this end of the continuum into what will be called Experimental Science Discourse. Again, the significant feature of the graph is that the percentage of experimental science topics and methodologies, as represented by the black bars, remains very close to the mean (45%) in the 1890 and 1900 volumes and from 1940 through 1970; that is, the amount of experimental science discourse in the volumes surveyed remains about the same during these periods—the same type of stability seen in Figure 4.2, the natural history progression. The sharp rise in interest in natural history during the early part of the century accounts for the corresponding decline in the presence of experimental science discourse during the same time pe-

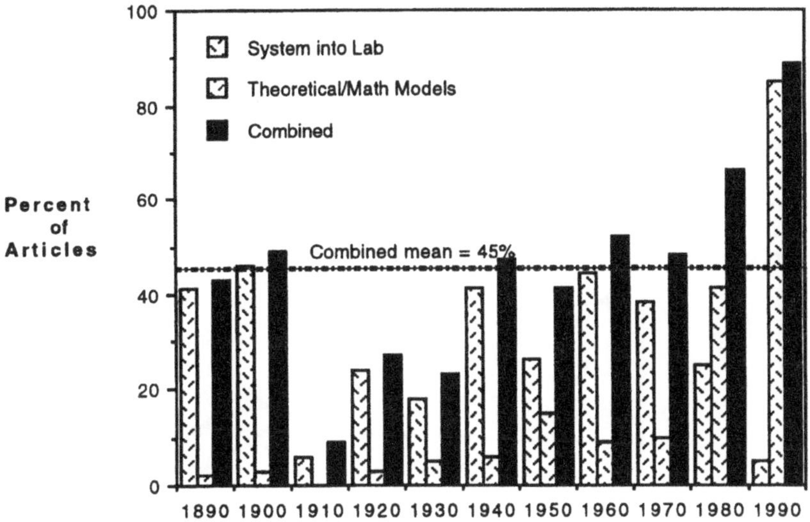

FIGURE 4.4. Experimental science discourse (N=217).

riod, whereas the surge in experimental science discourse after the 1970 volume corresponds to the rise of methodology using inferential statistics and computers.

The Transition From Natural History to Experimental Science

Having quantified graphically the two extremes of the continuum, it is now possible to combine them in order to see the interplay between natural history and experimental science in American ornithology. First, Figure 4.5 compares the two extremes of natural history and theoretical science. Seen here again is the clear predominance of field observation during much of American ornithology's history, with a sharp reversal only after the 1970 volume.

Now, combining the two categories of Natural History and Experimental Science Discourse created above in Figures 4.2 and 4.4, Figure 4.6 gives a more accurate picture of the interplay between natural history and experimental science discourse in American ornithology. What is significant for the present study are the volumes between 1940 and 1970. Though experimental science discourse is gaining ground during the period, natural history discourse holds its own. Only after the 1970 volume does natural history methodology and discourse give way.

But this may still be somewhat deceptive. Seven articles in the 1980 volume and 17 in 1990[5] are classified as theoretical simply because of the use of inferential statistics; yet they are primarily observation-based studies.

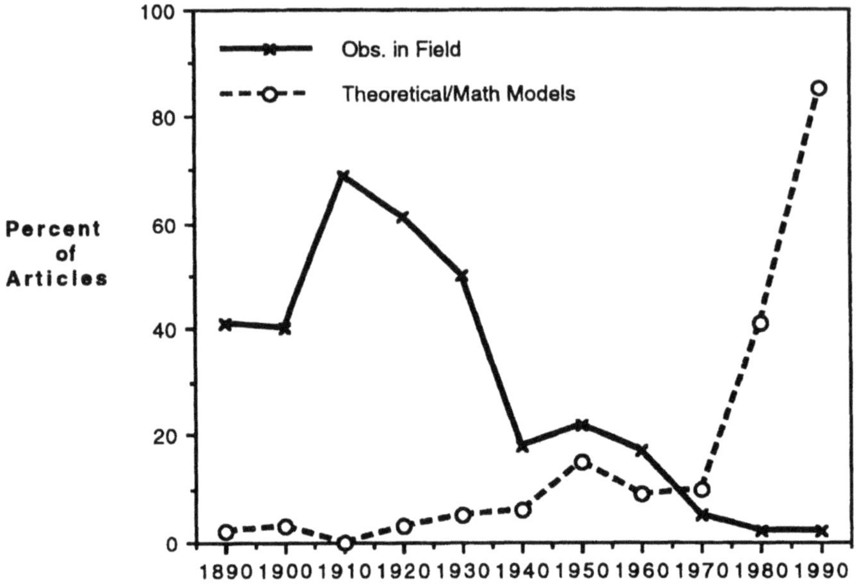

FIGURE 4.5. Natural history observation vs. theoretical science (N=215).

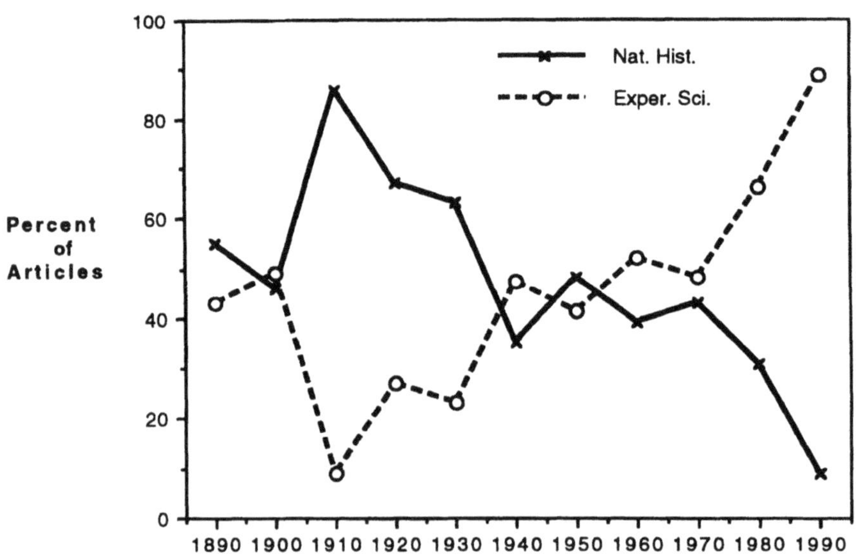

FIGURE 4.6. Natural history vs. experimental science discourse (N=412).

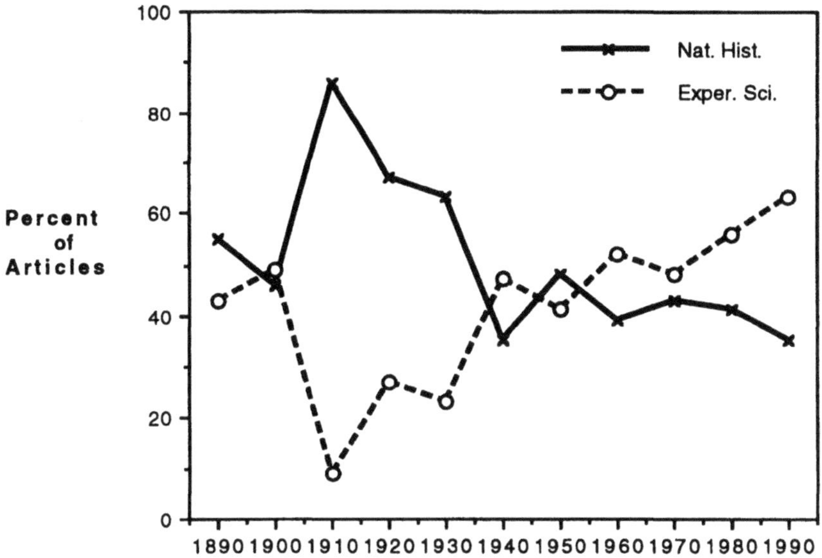

FIGURE 4.7. Natural history vs. experimental science discourse: Adjusted for 1980 and 1990 (N=412). Adapted from Battalio, 1996, p. 179. © 1996, Baywood Publishing Co., Inc.

Data analysis does not extend much beyond their immediate research hypotheses; and statistics seems to be added, sometimes only peripherally, for argumentative and, likely, publication purposes. For demonstration purposes here, Figure 4.7 transfers these 24 articles into natural history, where they perhaps more properly belong. This adjustment narrows the gap considerably in 1980 and 1990 between natural history and experimental science. Given that this realignment is appropriate, it can be said that only in the last 10 years has experimental science discourse in ornithology made inroads into natural history discourse. In short, it has taken about 100 years for natural history discourse to be displaced.

THE DISCOURSE SCHEMA REFINED

The insights provided by the Continuum of Scientific Perspectives on Nature allow us to refine the schematic diagram presented at the end of chapter 3. Figure 4.8 illustrates the revised discourse schema. First, because linguistic understanding, social factors, and personal experiences influence writers' *depth* of perspective, the poles of discourse are more appropriately symbolized synchronically as a *vertical* relationship, rather than a horizontal

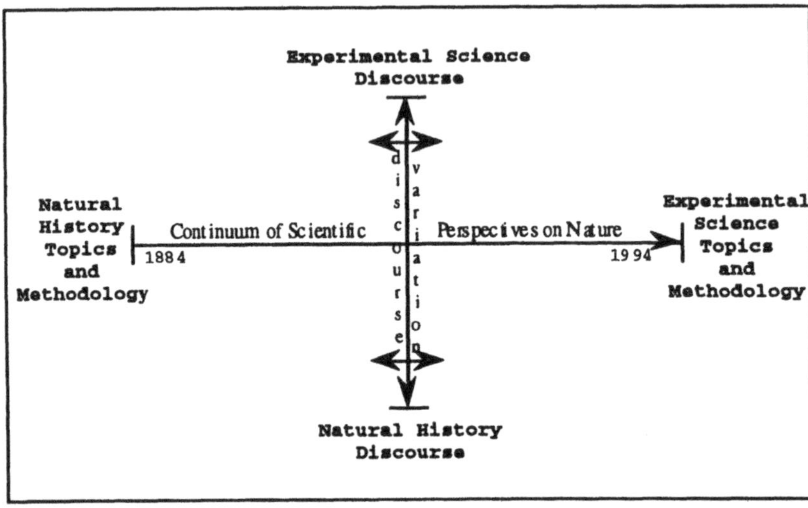

FIGURE 4.8. Discourse schema modified by diachronic, or historical, components.

one. However, placing experimental science discourse at the top of the discourse-variation line is not intended to represent its inherent superiority over natural history discourse, only that it becomes the dominant model once a science, like ornithology, professionalizes. Since the Continuum of Scientific Perspectives depicts a historical progression, it is more appropriately the horizontal component in the schema (that is, a time line beginning with natural history topics and methodology and evolving toward experimental science topics and methodology, reflecting writers' changing perspectives on nature). As the evolving Continuum of Scientific Perspectives passes through American ornithological time (1884 to present), the two poles of discourse adumbrate, as represented by the addition of arrows to the line of discourse variation. There are multiple arrows because writers are able not only to change their depth of perception but also to move back and forth on the Continuum, creating infinite forms of intermediate discourse. The resulting interplay of discourse creates the transitional discourse, which will be modeled in the rest of this chapter.

DISCOURSE OF TRANSITION: FROM AMATEUR TO PROFESSIONAL WRITING

As this chapter has shown, these two traditions of natural history and experimental science in American ornithology are not separate entities, but rather have arisen in conjunction with each other; one has influenced the other. At work here are not two stagnant poles of discourse functioning separately, but rather discourse gradually changing from one pole to the other, that is from natural history to experimental science discourse. As was pointed out in chapter 3, these models represent ideals that are not absolutely fixed, since scientific papers may actually represent trends in one direction or the other, rather than appearing as perfect examples of one discourse model or the other. Thus, a third discourse model is also appropriate for ornithology: a transition model that incorporates features of both natural history and experimental science discourse in varying degrees as the field of ornithology gradually changes from amateur to professional.

In developing the transition model, it is important to ask how the models for discourse change as ornithology advances. The answer to this question is tied to the kinds of systems excluded in scientific discourse; for, of its nature, scientific discourse leads to exclusion (see, for instance, Kuhn, 1970; Latour, 1987; Taylor, 1996). One system excluded is the *aesthetic*. Science does not deal with nature's beauty, since such judgments imply an invocation of subjectivity, which would undermine science's demand for objectivity. For similar reasons, *pathos* is not part of the presentation of scientific argumentation. Scientists deny the need to appeal to readers' emotions and feelings to prove an argument; scientific arguments rely instead upon demonstration via methods, statistics, and graphics. Science also does not hold *polemics* in high regard; rather than argumentation to degrade other scientists or their beliefs and principles, science chooses to disprove by refutation (Popper, 1959). Further, *political systems* are also excluded from scientific discourse. In modern times, there is a practical reason for this exclusion, since a primary source of funding for science comes from governmental agencies, a fact that requires science to remain free from political affiliation. Finally, *author presence* is not as evident in scientific writing. Traditional use, for instance, of passive voice attests to this preference; instead, objective presentation of method, procedure, and results supposedly guides the reader. None of these systems are absolutely excluded in scientific discourse, however. Any of them may appear in a given piece of scientific discourse due to varying circumstance, context, and audience. Thus, it is more proper to speak of a continuum connecting ornithological discourse models, rather than of mechanisms that isolate the two discourse entities.

Consequently, a transition discourse model is now presented, which enables writers to make the transition from natural history to experimental science discourse. As mentioned earlier, writers using both types of discourse will view themselves and their audience, hold worldviews, and use knowledge in various ways given audience, circumstance, and context. Thus, any given writer of transition discourse can be characterized by a unique combination of the essential premises already described in chapter 3 for each type of discourse. Therefore, a linguistic description of the discourse of transition can most easily characterize this third model—to which this study now turns.

A key figure during the transition in American ornithology from natural history to experimental science discourse was Frank M. Chapman, whose case study is presented below. The biographical sketch shows his appropriateness as a model writer of this hybrid discourse. Since Chapman had extensive experience writing for both the professional and amateur ornithologist throughout his career, a close look at examples of his writing for both types of audiences will reveal the major linguistic features of both types of discourse[6]. After comparing these features, we can then derive the linguistic components of a third discourse model, that of transitional discourse from amateur to professional ornithologist.

Frank M. Chapman: His Life, Work, and Impact on American Ornithology

Frank M. Chapman is perhaps the perfect choice as a transition figure in early American ornithology, one whose writings can be analyzed for features to be used in a discourse model representative of an early ornithologist writing for both amateur and professional. As is shown in the biographical description below, he communicated successfully with both amateur and professional worlds. He used his ornithological knowledge both to arouse public sympathies for birds and for their protection and to advance scientific knowledge of systematics and zoogeography. His worldview was both as a museum curator whose duty was to popularize birds through exhibition and as an ornithologist who wished to contribute to the advancement of his science. Most importantly, he, more than any other early American ornithologist, left an extensive *personal* record via his autobiographical works, as well as a scientific record, which can be studied and interpreted. Because Chapman published frequently in *The Auk*, the discourse of which is central to this book, the results of this case study also have direct bearing on the study as a whole. Because Chapman wrote *early* in the evolution of American ornithology, his work is also valuable as the basis for a transition model, the features of which will be a useful benchmark for observing changes in discourse in *The Auk* during the twentieth century as American ornithology professionalized. Since a goal of this study is to posit the features of discourse characteristic of the switch from amateur to professional writing, the model requires a subject who, like Chapman, wrote for an extensive period of time and whose writings demonstrate a wide range of features. Finally, since the transition model is identifying characteristic features of discourse as it changes from amateur to professional, it is important that the subject for study be someone like Chapman who had not been previously trained as a scientist and consequently would not have lost his or her amateur orientation. Thus, Chapman's writings are appropriate sources for developing a third discourse model, that of transition discourse.

One of the most influential American ornithologists from the turn of the century through the early 1920s, Frank Chapman was best known in ornithological circles for his *Handbook of Birds of Eastern North America* (1895), a standard bird identification guide of its time. In addition, he was one of the most popular writers of bird books, his popular book sales totaling 230,000 by 1933 (Mayr, 1975, p. 376). During his 54-year career with the American Museum of Natural History, Chapman published 17 books and 225 articles, which appeared in publications varying from professional journals like *The Auk*, to popular magazines such as *Natural History* and *National Geographic*. His interest in disseminating knowledge about birds and the natural world to the general public led him to publish frequently outside the scientific circle. Environmental concern for the welfare of birds prompted him to create *Bird-Lore*, which he convinced the fledgling Massachusetts Audubon Society to adopt as its own. As a result, his publication became the voice of conservationists, who rallied against the millinery trade. A brief look at his life will show the appropriateness of his selection as a transition writer whose discourse was of interest to both natural historians and experimental scientists.

The process involved in Chapman's decision to become a professional ornithologist exemplifies the socialization of a young enthusiast of birds. Though he began his career in 1880 in the collection department of a New York City bank at the age of 16, his casual interest in birds intensified greatly during this period as a result of his acquaintances with two other bird enthusiasts, Frederick Dixon, a fellow train commuter, and Clarence Riker, whose adventures in the Amazon where he collected 400 bird skins fascinated the young Chapman. During this time he also discovered John Burroughs' *Wake-Robin*, for which Chapman (1935) in his autobiography "[gave] gratitude for his [Burroughs'] aid in self-revelation . . . [since] I for the first time found expressed my own response to the song of the Wood Thrush" (pp. 28–29). In 1884, he read in *Forest and Stream*, the standard sportsman's magazine of the period, a request by the American Ornithologists' Union for volunteers to report on bird migration. The AOU's acceptance of his offer, Chapman says, determined his future; for ornithologist A. K. Fisher, impressed by Chapman's report, became his avid mentor. Through Fisher's sponsorships, Chapman joined the Linnean Society and in 1885 became an associate member of the AOU. In addition to these contacts, Chapman's purchase of Coues' ornithology textbook *Key to North American Birds* made him aware that contemporary studies of birds were being conducted. Thus, in 1886 Chapman (1935), at the disbelief of family and coworkers, quit his respectable and well-paying bank position, deciding instead to become a "professional ornithologist. I had no plan other than to make a collection of the birds of the vicinity and record my daily observations on them with the ultimate object of publishing the results in the form of an annotated 'local list'" (p. 45).

In 1887, while in Florida pursuing these objectives at his family's summer home, he met ornithologist George B. Sennett, who maintained a private bird collection at the American Museum of Natural History in New York. Seeing Chapman's enthusiasm and ability, Sennett obtained permission for Chapman to bring his collection of Florida specimens to the museum and hired him to help care for his own collection. While at the museum, Chapman volunteered to assist Curator J. A. Allen catalog a major collection of recently purchased South American specimens. The next year, Allen, also AOU president and editor of *The Auk*, offered Chapman a position as his assistant at the American Museum, where he remained for the rest of his long career.

The position, which he obtained through personal contacts rather than scientific competence, enabled him to rise quickly in ornithological circles. He became associate editor of *The Auk* from 1894 to 1911. His collecting expeditions for the museum carried him to all parts of both American continents—to the Caribbean, West Indies, and Mexico early in his career collecting specimens and acquiring both personal knowledge and museum expertise; to South America studying geographical distribution; and later to the Canal Zone observing the habits of tropical species. In 1901, he was named Associate Curator of Birds and Mammals, becoming in 1908 the Curator of Birds exclusively. He became president of the AOU in 1912 and an honorary member of its British counterpart in 1917, receiving his Doctor of Science from Brown University in 1913. He lectured all his life on birds, accepting invitations to speak across the United States and in Europe as well. In the 1920s, he continued his association with bird protectionists as chair of the Board of Directors of the National Association of Audubon Societies.

During his career he was amply awarded for his accomplishments. From the National Academy of Sciences he received an award for his volume on the zonal distribution of the birds of Colombia and in 1928, along with Lindbergh, was awarded the Roosevelt Medal for "distinguished service in the promotion of the study of natural history . . . to [one] who has taught a nation to see, to know, to love and to protect [birds] . . . a creative innovator in methods of exhibition and of ornithological research" (as cited in Chapman, 1935, p. 331 facing). By his retirement, he had transformed the American Museum into an international center for bird study, both by his own South American collecting and by addition of other museum collections, particularly in 1932 of the prestigious British Tring collection of 280,000 specimens.

Chapman's obituarist, co-worker Robert Cushman Murphy (1950), described well the double role of both amateur naturalist and professional scientist that Chapman successfully juggled throughout his career:

As a writer and lecturer, he won early reputation of being the most articulate ornithologist of his generation. Because he was as much concerned with the habits and protection of living birds as with their classification and relationships, he exerted wide influence in creating popular interest. (p. 312)

The various titles Chapman gave to himself in his autobiography[7] indicate this duality as he attempted his own self-identity. He described himself on the one hand as a bird-lover and nature-lover, who had often read and re-read with fascination a popular book describing the natural beauty of Trinidad and who was struck by Burroughs' description of a bird's song. On the other hand, in describing his theories on bird distribution and empirical contributions to systematic ornithology, he likewise named himself a naturalist, a field ornithologist, and a faunal naturalist. Thus, in Chapman's own mind, he wore many hats.

As a model for amateur naturalists, as well as their associate, Chapman felt the need to popularize ornithology; otherwise, the movement to protect birds would be futile. He was involved in this popularization throughout his life by working on behalf of the fledgling National Audubon Society and by offering his services as editor of their national publication. He lectured to children's groups, women's clubs, and educational institutions throughout the country on the need for bird protection and often wrote for general audiences, as for example his series in the New York *Evening Post* on the birds of Central Park. His popular books on birds included *Bird-Life: A Guide to the Study of our Common Birds* (1897) and most notably his *Handbook*, which he acknowledged was written to reach a wider audience than just ornithologists. As a result, the book was in print for over 30 years. Through his innovative work in museum exhibition, he hoped to excite visitors' interest in nature. Thus, he considered the museum curator's function to be primarily educational: "His success is to be measured by the extent to which he arouses in his unseen audience a desire for more information than can be supplied by exhibitions, labels and guidebooks" (1935, p. 71). There was likewise an element of the transcendent in Chapman's attitude toward birds. Like Thoreau and Burroughs, he was awestruck by bird song, commenting on numerous occasions about the beauty of nightingales, thrushes, and the alluring, tropical "poor-me-one." Like them, he described the struggle between the naturalist and the scientist over specimen collecting. Like them, he kept a journal of his ornithological experiences, which resounds with emotional expressions derived from both his observations of natural beauty and his closeness to nature, at one point including a Thoreauvian admonition to live in the present. Typical of this metaphysical perspective was his description of his calling to ornithology:

Beyond the birds' universal appeal of "form, song and rhythmic flight" there is, for the elect, something "far more deeply intertwined," some rare and

> precious heritage from ancestors to whose untutored minds birds were signs and symbols of the mysterious manifestations of nature. (1935, p. 6)

He likewise closed his autobiography on a similar note, commenting that, just as robins' songs "arouse memories of a thousand Robins singing down the pathway of the years I share a span of their immortality" (1935, p. 384).

Juxtaposed against this perspective is Chapman the professional. The scientific theories and taxonomic dilemmas he researched throughout his career affirm that he derived more than emotional satisfaction from the study of birds. His museum expeditions focused on research in both island and altitudinal distribution of birds. In the process, he added to the taxonomy of birds and mammals of the world through his discovery of many new species. He broadened ornithology's influence, interconnecting the study of birds with geology and geography by mapping four faunal life zones in the Andes region. Chapman was likewise an innovator in methodology, demonstrating the value of still and motion-picture photography in documenting birds' habitats and behavior. He published professionally throughout his lifetime, presented papers at every AOU meeting for over 40 years, and accepted prestigious speaking invitations in Europe, such as those to the British Zoological Society, the International Ornithological Congress, and the Royal Geographical Society. His awareness of the scientific community of which he was a part was evident both by his continuing correspondence with South American collectors and museum curators and by his dutiful attendance at both national and international ornithological meetings, which, he acknowledged, enabled him to renew professional acquaintances. From his worldview as a professional, the title "ornithologist-curator" gave its owner special power and authority to advance his expeditions, gain access to restricted areas, and meet both royalty and leaders of government. From a scientist's perspective, Chapman (1935) defined scientific objectivity through two guiding principles: in the museum "[s]cientist, artist, and *preparateur* must collaborate harmoniously in its production, but with the first-named always in control. Never must truth be sacrificed to beauty" (p. 167); and in the field "[t]here are occasions when the 'bird in the bush' is supreme, but only the bird in the hand will satisfy that desire for intimate, exact knowledge of its characters which obsesses the naturalist" (1935, p. 49).

Linguistic Features of Transitional Discourse

Four examples of Chapman's writing, two each written for amateurs and professionals, respectively, were chosen for this analysis. The methodology follows the same procedures and identifies the same 32 linguistic features described in chapter 3 for developing models for natural history and experimental science discourse. The purpose of this chapter's analysis is to identify two sets of linguistic features, one set for amateur and one for professional, employed by the same writer writing for both audiences. Comparing these two sets of features with the features identified in chapter 3 for natural history and experimental science discourse should provide a general set of rules for evaluating the processes involved in discourse as it transitions from amateur to professional, from natural history to experimental science. As explained above, the writings of Frank Chapman are fitting subjects for this analysis because of Chapman's status within American ornithology, his lifelong presence in both amateur and professional contexts, the diversity and volume of his publications, and his presence in early American ornithology.

Selecting articles published within a 6-year period helps to assure that the linguistic differences identified in the four articles are more likely associated with Chapman's decisions as a writer for amateur and professional, rather than with either social, historical, or institutional changes or an evolution of his personal writing style. There would have been little, if any, influence on his writing style by peer review in either of the two publications in which the selected articles appeared. Chapman himself was the editor of *Bird-Lore*, the popular magazine used in the analysis. As will be explained in chapter 5, *The Auk* at this time was not refereed and regularly printed all articles submitted to it.

It may be argued that a larger sample of writing or additional authors should have been used, but it should be noted again here the impossibility of separating completely an individual writer's influence from any piece of discourse, regardless of the number of authors or samples of writing. Since past experiences, objective norms established by the discourse, and the audience itself influence the decisions writers make, there will always be some variance in the empirical results of such a study since individual variation is inherent in all writing. In addition, the purpose of this portion of the study is to develop a general model of discourse, which can be compared with the other discourse models, as well as with a larger sampling of discourse from *The Auk*. It will be shown in later chapters that the discourse models, though based on a very limited sampling of discourse, mesh well with American ornithological discourse history, the linguistic samples from *The Auk*, and the discourse schema.

The articles

The two articles representative of Chapman the popular writer were taken from the magazine *Bird-Lore*, published by the National Audubon Society. In the first issue of the magazine, Chapman (1899), also the editor, described well its purpose and target audience:

> With this phenomenal and steadily increasing interest in bird-studies, there has arisen a widespread demand for a popular journal of ornithology which should be addressed to observers rather than to collectors of birds, or, in short, to those who study "birds through an opera-glass."... It is to supply this want [for communication] of bird students and bird protectors that BIRD-LORE has been established.... We would appeal to all bird-lovers for such support as we may be deemed worthy to receive. (p. 28)

The first popular article analyzed is "Roseate Spoonbill," published in *Bird-Lore* in 1914. Chapman's purpose in the article was to give readers a look into the life history of the Roseate Spoonbill, a water bird found primarily along the Gulf Coast. Before doing this, however, he gave readers a brief overview of the spoonbill's plight at the hands of plume hunters, which allowed him to discuss the Audubon Society's role. He then described the birds' ornithological history, not by scientific description, but by showing how the American spoonbills had become distributed throughout the world. As he discussed the birds' American range, his reference to Florida gave him another opportunity to discuss the Audubon Society's attempts to protect Florida birds. Instead of a textbook-style life history of the spoonbill, Chapman recounted an adventurous night he spent in a spoonbill rookery. While describing what he saw that night, he interwove the birds' life history, discussing their nesting cycle, raising of young, and roosting behavior, often in anthropomorphic terms.

"Notes from a Traveler in the Tropics: I. Down the Coastline to Cuba," the second popular article analyzed, appeared in the 1918 *Bird-Lore*. It was part of a continuing series of articles on a recent trip Chapman had made to South America. In the first part of the article, he described the habitat and named the birds he saw from his train as it traveled along the Florida coast to the Florida Keys, where he boarded a boat to Cuba. He then recounted the number and kind of birds he saw during his short stay at the Isle of Pines on the southern coast of Cuba, ending with a short description of the bird song he encountered on the Isle. In this popular article, he merely named the birds he saw and did not discuss their physical features, except to comment occasionally on a bird's beauty. However, when naming Cuban species, he did add a genus name in parentheses, it can be assumed, to help amateur readers locate these unfamiliar species in a field guide.

"The More Northern Species of the Genus *Scytalopus* Gould," which appeared in the 1915 *Auk*, is the first example of professional discourse. Here, Chapman suggested a revision of a genus of wren-like birds from the Andes, where he made frequent collecting trips during his career. After discussing the difficulty of collecting such small species in the dense subtropical jungle, he recounted the origin of the over 133 specimens he used in his study. He then listed his suggested revisions to the taxonomy of the genus and discussed the species' ranges. The rest of the article consisted of mostly full-page descriptions of each of the eleven species under discussion. Each description explained the species' general characteristics, type, range, and plumage. The article ended with a key to the adults of the species and a synopsis of the characteristics of some juvenile birds.

The second professional article by Chapman is another taxonomic description, this time a "Description of a Proposed New Race of the Killdeer from the Coast of Peru" published in the 1920 *Auk*. The style and content parallel Chapman's article above on the Genus *Scytalopus* Gould. He first described the specimens found, noted the species' range, and then gave a detailed description of the plumage features for both adult and young of the subspecies. Typical of modern-day research articles, he ended with a call for more research.

The analysis

Table 4.2 summarizes the linguistic analysis of the selected works by Frank Chapman. Totals for each linguistic feature were obtained by averaging the occurrences in the two articles for amateur and two for professional writing, respectively. Chapman's discourse is clearly shown to be discourse in transition because it incorporates many features distinguishing both discourse models presented in chapter 3, as well as showing gray areas where the discourse is still in the process of evolving.

The following eight features, proposed in this chapter as features whose number of occurrences indicate writers whose styles reflect either a natural history or experimental science discourse model, had likewise become delineated in Chapman's writing. It is therefore suggested here that the frequent occurrence of these features will be early evidence that discourse has evolved from natural history to experimental science:

1. an increase in average sentence length[8]
2. an increase in nominalizations
3. appearance of technical and polynomial technical terms
4. less emphasis on subjects as place-holders
5. decreased use of the verb *to be*
6. a decrease in the number of simple sentences
7. an increase in the number of subject-verb interruptions
8. an emphasis on reliability of evidence

The second set of features listed below are those that are not characteristic of Chapman's scientific writing, though they appear in the experimental science discourse model described in chapter 3. Since Chapman was writing early in the evolution of professional American ornithology, the eight features below are therefore suggested here as features that are later to appear

TABLE 4.2.
Linguistic Analysis: Transition Discourse

Linguistic feature	Amateur	Professional
Average sentence length	29.3	33.4
Metadiscourse		
refer to own thinking	2	2
refer to structure of writing	1	0
refer to reader's act of reading	0	1
Nominalizations	6	13
Technical terms	2	12
polynomial technical terms	0	7
Personal pronouns	17	9
Use of definitions	1	1
Sentence complexity		
subjects as place-holders	11	4
passive voice	8	9
to be	32	17
reporting verbs	4	1
active verbs	43	39
% simple sentences	26	27
Complex clauses		
relative noun	15	15
relative adjectival	11	7
time	6	3
causal	1	3
place	2	2
conditional	10	2
Subject-verb interruptions		
% interruptions	27	30
number of interrupting words	128	122
Evidentials		
reliability of evidence	10	14
belief	1	1
induction	1	
hearsay	1	0
citation	0	2
deduction	6	0
hedging	1	1
sensory evidence	7	1
expectations	17	15

in the transition to experimental science discourse. Consequently, once these features are identified, the transition to experimental science can be considered completed:

1. increased use of metadiscourse
2. minimal use of personal pronouns
3. increased use of definitions[9]
4. preference for passive voice
5. increased use of simple sentences
6. restricted use of complex clauses, though an increasing use of causal clauses reflects the change from natural history to experimental science discourse
7. restricted use of evidentials
8. a striking increase in citations

Format and Structure of Transitional Discourse

Typical of natural history discourse, Chapman's articles in the Audubon Society's *Bird-Lore* include drawings of birds in their natural habitats. In the "Roseate Spoonbill," Chapman inserted a color painting of the birds facing the title page to his article. This was one of the few color pages in the issue, certain to attract readers' attention. Rather than selecting a close-up of one spoonbill for readers' detailed inspection, Chapman instead selected a scene with many roseates feeding together along water's edge. The association of the birds with their surroundings was a particularly appropriate selection, since it reflected the developing interest of the Audubon Society in the conservation of birds and their environment. Chapman's other visual was a space-filler at the conclusion of his article, a drawing that he, as editor, likely chose. It shows three perched birds in springtime, an emotionally appealing drawing to the bird watchers reading his article. "Notes from a Traveler in the Tropics" includes two half-page drawings of birds perched in their habitats, one of frigatebirds and another of anis, two species named in his article. Chapman's choice of visuals in both articles reflects the function visuals perform for the natural history writer (that is, as an emotive rather than intellectual aid). Unlike Chapman's *Bird-Lore* articles, his scientific articles include headings, introducing by taxonomic name each of the species being considered. Within each description, he subheaded each avian feature discussed. His "*Scytalopus* Gould" article also included simple graphics: two short lists of scientific nomenclature, three tables of specimen measurements, and two outlines of the major parts of his argument.

REFINING THE DISCOURSE MODELS : AN ANALYSIS OF DISCOURSE IN PRACTICE

The linguistic analyses used to delineate the three discourse models presented in chapters 3 and 4 are based upon a few samples from representative authors. Consequently, there is a need to analyze a larger sampling of ornithologist-authors writing over an extended period of time in order to test the premises and linguistic features proposed in this study. Discourse from American ornithology's professional journal, *The Auk*, is again the most fitting source for such an analysis. Thus, in chapter 5, after a historical overview of changes in topics and methodology in American ornithology, articles written by 44 authors, four per 10-year periods from 1885 through 1990, are sampled and analyzed using the same criteria and coding form previously described. Once the discourse models have been refined, chapter 6 places this study of American ornithological discourse in context with other contemporary studies of scientific writing, thus enhancing our understanding of scientific discourse.

5

A Hundred Years of American Ornithological Discourse

As shown in chapter 2, the science, and consequently the discourse, of American ornithology was still in its infancy in the nineteenth century, having emerged from its roots in the ornithology of the seventeenth and eighteenth centuries with the founding of the American Ornithologists' Union in 1883. Writing in the Essex Institute Proceedings in 1856, Massachusetts writer Samuel P. Fowler summed up well the state of mid-nineteenth-century American ornithology:

> It is greatly to be regretted, that we have not at this time, a cheap and complete work, with specific descriptions of all our birds, as it would very much facilitate the study of ornithology, and increase the number of those, who would like to learn more of their history, and although many valuable and beautiful books have appeared upon the science of ornithology, still it would be difficult to direct an enquirer where he could obtain a work, describing our feathered tribes at a moderate price. (Fowler, 1974/1856, p. 334)

Professionally, the only Americans whom ornithological historians list among the ranks of internationally respected ornithologists by 1850 were artist John James Audubon (Farber, 1982, p. 104) and scientist John Cassin (Kastner, 1986, p. 14).

Not until the 1880s was there an attempt to organize and systematize data on migration, geographical distribution, classification, and nomenclature of

American birds (Mayr, 1951/1975, pp. 378–380). Reference guides confused the names of birds and often contained poor illustrations. Not understanding a bird's life history, authors sometimes identified immature birds as separate species. Even accounts by the masters Wilson and Audubon contained errors. For instance, the appendix to volume 5 of Audubon's *Ornithological Biography* acknowledged "errors relative to the species" (1831–1849, p. 337). Wilson erroneously pictured chimney swifts as perching in trees. As for ornithological expertise in general, no one yet understood completely the many plumage variations and molts. Because ornithological science was still developing, early twentieth-century ornithology still relied on amateur publications. As late as 1949, ornithologist Ludlow Griscom combined Thoreau's journal accounts of bird migration and distribution with other Concord ornithologists' observations to produce *The Birds of Concord*, "the first comprehensive study of the changing bird populations of an area over an entire century" (Harding, 1981, p. 48).

The development is not a continuous one, nor one with a clear focus; rather it is an irregular patchwork of both attempts to systematize the relationships among birds and presentations expounding upon their beauty, intrinsic worth, and natural history. For systematists, avian classification was a continual problem since birds were being discovered, sometimes in the hundreds during the eighteenth and nineteenth centuries, by explorers and naturalists throughout the world. For nature writers, mythologies and superstitions about birds persisted because accurate, realistic presentations of the natural history of birds required many years of study before a bird's life cycle was fully understood.

As shown by the graphics at the beginning of chapter 4 depicting this progression of American ornithology, only during the past two decades has the discourse of experimental science superseded that of natural history. The persistence of both discourses over so long a time justifies both the presentation in chapters 3 and 4 of three corresponding discourse models in ornithology—natural history, experimental science, and transition models—and here a closer analysis of the discourse itself as found in *The Auk*. This analysis will provide the basis for a refinement in chapter 6 of the discourse models already presented.

HISTORICAL CHANGE IN AMERICAN ORNITHOLOGICAL DISCOURSE

This closer analysis will again focus on chapter 4's database of 444 articles from *The Auk*, but now in much more detail. Studying changes in methodology and topics for research over the past century in conjunction with corresponding changes in the discourse of ornithology will shed light on the persistence of natural history (that is, as defined here, of observation-based study carried out by amateurs), even into 1990. Based on chapter 4's time lines charting the ornithological Continuum of Scientific Perspectives on Nature, ornithological history may now be broken into four periods (see Figure 5.1):

1884–1900: The Discourses of Natural History and Experimental Science in Balance
1900–1940: The Predominance of Natural History Discourse
1940–1970: An Interplay of Discourse
1970–1990: The Emergence of Experimental Science Discourse

The following survey of ornithologists' discourse in *The Auk* during each of these four periods will demonstrate the gradual change from amateur to professional discourse. The survey will describe in detail the content of the 11 volumes already used in chapter 4's 100-year survey of methodology and topics in American ornithology. Specific examples from the 444 articles constituting these 11 volumes of the journal will show the delayed movement along the ornithological discourse continuum from natural history to experimental science discourse, a movement already discovered during the previous chapter's analysis. Summaries of the evolution of organization,

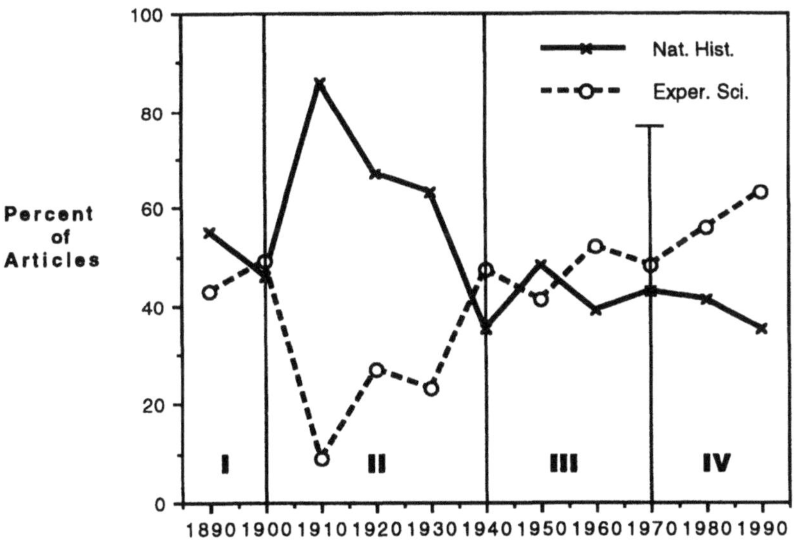

FIGURE 5.1. Four periods of American ornithological discourse. Adapted from Battalio, 1996, p. 179. © 1996, Baywood Publishing Co., Inc.

formatting, and graphics over the four ornithological periods will also be presented. The chapter ends with a linguistic analysis of a subsample of articles from the 11 volumes.

It should be noted here that the focus of the following survey is on the *discourse* of American ornithologists. Categorizing a particular article as natural history or experimental science discourse is not meant to be an evaluation of the scientific contribution itself. On the contrary, many scientific contributions to ornithology have been made over the years by amateurs. Similarly, the fact that a professional ornithologist has written in a natural-history narrative style does not negate either the worth of the contribution or his or her professional competence. The survey is merely documenting changes in the discourse itself as ornithological science progresses. It is not evaluating the value of the scientific information described by the discourse.

1884–1900: The Discourses of Natural History and Experimental Science in Balance

On September 26, 1883, 21 of 48 invited American ornithologists met at the library of the American Museum of Natural History in New York City to elect officers and convene the first meeting of the American Ornithologists' Union. Coming from all walks of life, this first nucleus of American ornithologists represented both the technical and the amateur. As shown by an analysis of the 1890 and 1900 volumes of *The Auk*, these members' ornithological research maintained a balance between laboratory and field work, between specimen studies and natural history accounts.

1890

Of the 42 articles published in the four issues of 1890, more than half (23) concerned the natural history of birds. The articles were essentially personal accounts of topics such as robin roosting, migration notes, winter distribution of Bobolinks, and nesting habits of numerous species. The following excerpt of a description by Frank Bolles (1890) of the mobbing of Barred Owls is typical of the narrative writing of this early period:

> Whenever on my summer walks I came to a spot which I wished to "sample" for its birds, I would place Puffy [the author's captive Barred Owl] on a bending sapling, and hiding in the neighboring foliage, I would "squeak" by drawing in my breath over the back of my hand, and attract the attention of any birds which were near by. Usually in the deep woods the first comer was a Red-eyed Vireo, Chickadee, Hermit Thrush, or Oven-bird; but whichever it chanced to be, an alarm was almost sure to be given that would bring birds from all directions eager to see the cause of disturbance. Even when I was

> imperfectly concealed, the irritated crowd paid little attention to me, provided I kept reasonably quiet. (pp. 108–109)

There were also literary flights of prose, such as this description by O. Widman (1895) of winter robin flocking:

> Now they begin to feed, running swiftly over the newly burnt ground, gathering food at every run, when suddenly they spy a large body of others of their kind, coming nearer and nearer, in a long-stretched line, filling the air with a medley of melodious whistles in many different keys—and up they go like a flash to join their passing brothers. The whole troop, perhaps 500 in all, maneuvers now in common and like a regiment of the best drilled soldiers, they perform the most astonishing evolutions in turns and sweeps, now high, now low, now all a flash of brightest gold, then all a streak of silvery white, almost vanishing from view in the distant sky, to return with lightning rapidity so low as to almost touch the ground with the tip of their long, swift pinions. (p. 8)

Seventeen articles were primarily personal annotated bird lists of areas such as Florida, Arizona, and Brazil. Several lists, such as those for Cape Cod, Massachusetts; Oneida County, New York; and the Dry Tortugas, west of Key West, Florida, also included lists of specimens taken. Figure 5.2 shows a page from a typical article of this type, here of an annotated list of Pickens County, South Carolina, which includes a long narrative description under entry 10 and a more technical description in entry 14 giving the measurements of a specimen taken in the field.

The journal also included numerous technical studies based on an examination of carefully prepared study skins obtained from individual bird specimens collected in the field. Of the 17 articles in 1890 that contained detailed descriptions of collected specimens, seven proposed new species or subspecies, including descriptions of new species of wren, vireo, junco, swallow, hummingbird, blackbird, and wild turkey. There was, however, only one article that could be classified as theoretical, an argumentative criticism by editor J. A. Allen (1890), of the naming of some geographical forms of North American birds. Here, the writing style and tone are more authoritative than Bolles' excerpt above:

> Variations, even when slight, are facts of great import, whether they be individual, climatic, or specific, in the usually accepted sense of this last term; and when, except in the case of purely individual variation, they are made the basis of a name their availability in the discussion of profounder questions is thereby greatly enhanced. Down to as late as 1865, a collection of individuals indicated by a binomial name was a species, theoretically, at least, distinct and definable from any other similar group. An occasional intermediate, when

1890.] LOOMIS, *Summer Birds of Pickens County, S. C.* 37

10. **Falco peregrinus anatum.** DUCK HAWK. 'SPARROW EAGLE.'— As my guide and myself were picking our way in a dense fog along the broken trail at the foot of the precipice at Table Rock on the 15th of June, three Duck Hawks with loud cries bore down upon us from the mist-hidden crag above. They flew about in a manner that plainly indicated that our presence filled them with anxiety. Occasionally one would alight for a few moments on a dead tree. Several shots were fired, but the towering cliff led me to miscalculate the distance, and the only effect produced was to frighten them away. In about half an hour they re-appeared, screaming as before. In the meantime the fog lifted and the surface of the cliff became distinctly visible. The object of their solicitude was soon apparent, for one of them, with a ground squirrel in its talons, alighted in a crevice in the massive wall of rock. Several unsuccessful attempts were made to obtain a specimen, which resulted finally in driving them from the scene. Four days later I visited the spot a second time hoping to secure one with a rifle, but they had grown extremely shy, remaining out of sight at the top of the precipice. On the 22d I rode across the range from Mt. Pinnacle to Table Rock and, when the brink of the cliff was reached, the three Falcons were again encountered, but the efforts put forth to capture one were as unavailing as were those on the former occasions. I did not find this bird elsewhere in my excursions about the mountains.

I was told that the Bald Eagle in the past bred at Table Rock. There is a great hole—affirmed to have been formerly a nesting site—in an inaccessible part of the cliff which still bears the name of 'the eagle hole.' The presence of an Eagle now, at any season, is an exceptional occurrence, and is of sufficient interest to awaken general comment. The Golden Eagle probably occurs during winter, if not at other times of the year. The Eagle, however, spoken of by my mountain friends is asserted to be a white-headed one.

11. **Falco sparverius.** AMERICAN SPARROW HAWK.—So far as my observation extended, the Sparrow Hawk was not numerous.

12. **Coccyzus americanus.** YELLOW-BILLED CUCKOO. — Tolerably common. Familiarly known throughout the region as the 'Rain Crow.'

13. **Ceryle alcyon.** BELTED KINGFISHER. 'KINGFISHER.'—One was seen July 9, 1886, along the Oolenoy at the bottom of Mt. Pinnacle.

14. **Dryobates villosus audubonii.** SOUTHERN HAIRY WOODPECKER.—Although the Hairy Woodpecker is common and generally dispersed, I did not satisfactorily ascertain whether other than the southern form was present. A female, shot June 6, 1888, at an altitude somewhat above 3000 feet measured in length, 9 in., tail, 3.4 in., dimensions that are indicative of the typical *D. villosus*, but the wing measurement (chord) of this example was only 4.6 inches.

15. **Dryobates pubescens.** DOWNY WOODPECKER.— Observed from the valleys up the mountain sides to their summits; moderately common.

16. **Ceophlœus pileatus.** PILEATED WOODPECKER. 'WOOD HEN,' etc.—For so large a bird, common. Generally distributed. Though ordi-

FIGURE 5.2. Annotated bird list from the first period of American ornithological discourse (Loomis, 1890, p. 37). Used with the permission of the American Ornithologists' Union.

> found, was a great stumbling block, usually to be accounted for on the theory of hybridization. Varieties, in the geographical sense of today, were rare indeed, and the term subspecies was almost unknown in zoölogy. (p. 3)

Thus, in 1890 the journal was heavily oriented toward natural history discourse, written in large part by the amateur.

1900

In 1900, the balance between natural history accounts and specimen studies remained essentially the same. There were personal accounts describing the natural history of species such as Black Rail, Rufous Hummingbird, and Ivory-billed Woodpecker, as well as the nesting habits of Cerulean Warbler and Franklin's Gull. One of these articles, "Notes on the Breeding Habits of the American Golden-eyed Duck or Whistler," reads like a Thoreauvian journal:

> June 8. I visited the Whistler's nest shortly before daybreak this morning, approaching it with great caution. The old bird was absent and at first I feared that she had removed her young during the night, but on looking into the hole I was delighted to find them still there, huddled closely together in a circle, and shivering a little, for the air was keen. There were, however, only *six* of them with the two eggs still unhatched. What can have become of the remaining three eggs or young? (Brewster, 1900, p. 211)

Personal bird lists also continued, as for instance from Idaho; Washington; and Refugio County, Texas, showing that American ornithology had become truly national in scope. However, the inclusion of an article on American ornithological history, discussing the dating of an eighteenth-century tract on American birds, suggests a new science still in search of its identity.

On the more technical end, 16 specimen-skin studies were included in 1900, eight of these proposing additions or changes to the taxonomic list, including new species of wren, jay, sparrow, shearwater, and three new Alaskan birds. Figure 5.3 shows a page from an article representative of nineteenth-century "technical" ornithology, describing a specimen of an Alaska Grouse by giving its type designation, measurements, and a technical description.

Again, as in 1890, there was only a single theoretical article, one which opposed Directive Coloration Theory, a theory emphasizing the role of evolution in the coloration of bird plumages. Included also was a defense by editor J. A. Allen (1900) of his analysis of plumage variations, a polemic regarded today by scientists as unprofessional:

> A careful study of Mr. Birtwell's paper has convinced me that there is nothing very unusual about Mr. Birtwell's caged Indigo Bunting. No "hypothesis" is

114 BISHOP, *New Birds from Alaska.* [Auk April

Canachites canadensis osgoodi, subsp. nov. ALASKA GROUSE.

Type, No. 4310, Coll. of L. B. Bishop, ♀ ad., Lake Marsh, Northwest Territory, July 5, 1899; (W. H. Osgood) L. B. Bishop.

Subspecific characters. — Similar to *Canachites canadensis* but with the ochraceous buff bars replaced everywhere by cream-buff and grayish white. On the upper parts the gray tips are paler, the ochraceous buff replaced by cream-buff and whitish, and the pale bars of the cervix grayish white instead of buff; below the white tips are larger, the pale bars whitish and cream color instead of buff, becoming cream-buff only on the jugulum.

Distribution. — Northwest Territory, Northern British Columbia and Alaska north of the coast mountains.

Measurements of type. — Length, 15.50; extent, 24.25; wing, 7.20; tail, 4.54; tarsus, 1.32 inches. Average length of 4 ♀, 15.38 inches. Average of 6 ♀ and 1 ♂ in wing, 7.02; tail, 4.51; tarsus, 1.34 inches.

Description. — Above irregularly barred with grayish white and black, the pale bars becoming cream-buff on crown and interscapulars and grayish buff on wing-coverts and concealed bars of rump and upper tail-coverts; scapulars and inner tertiaries black, irregularly barred on outer web and tipped with cream-buff and gray, with central wedge of white. Quills dusky, mottled on outer web and obscurely tipped with whitish; tail dusky, vermiculated, chiefly on outer web, with grayish buff and tipped with tawny ochraceous. Below barred with grayish white and black, the feathers especially on breast and abdomen broadly tipped with white; the pale bars becoming cream-color on throat and toward base of feathers on sides of breast, and distinctly cream-buff only on jugulum; flanks vermiculated with black, grayish white and pale cream-buff with broad central wedge of white. Tarsal feathering pale mars brown obscurely spotted with darker.

Remarks. — In worn breeding plumage adult females of *osgoodi* differ from *canadensis* from Maine as described above. After completing the summer moult and in early spring females from Alaska differ from females from Ontario and Quebec in the same manner but to a less degree, having the buff everywhere, especially on the cervix and abdomen, paler and the white tips below broader.

A summer male of *osgoodi* from Thirty Mile River differs from summer males of *canadensis* and *labradorius* only in having the tail tipped with paler rufous.

Two adult females of *labradorius*[1] from Mr. Bangs's collection

---

[1] Mr. Bangs writes me that these are not the best examples of *labradorius*, but are the most characteristic at present available.

FIGURE 5.3. Technical ornithology (specimen study) in the first period of American ornithological discourse (Bishop, 1900, p. 114). Used with the permission of the American Ornithologists' Union.

> necessary to account for its moulting at the time it did, nor would Mr. Birtwell have thought any "hypothesis" necessary if he had known the simple facts that the species normally moults twice a year. . . .
>
> Since Mr. Bonhote finds that I have "adduced no proofs in favour of non-colour change" in my paper on this subject published in 1896, it is hardly worth while to discuss the subject further in the present connection. In view of such a statement, however, I can hardly believe that my critic has given the paper in question very careful attention. As the subject is at present receiving renewed consideration, I am quite willing to await the results of expert investigators in this special line of research. . . . (pp. 331, 336)

These personal attacks by the journal's editor are further evidence that nineteenth-century American ornithology had not as yet perceived itself as a unified body of professionals. Though Greg Myers (1990) documented a contemporary scientific controversy waged by biologists in print (pp. 101–140), he noted that the "usual method of dealing with research claims one thinks are wrong is to ignore them; if they are not picked up by anyone, they will disappear into the morass of scientific publications" (p. 101).

Organization and formatting

In this early stage of ornithological discourse, there was no attempt to establish a standardized format, though there were individual paragraphs in the running text that were implicitly introductions and descriptions of method, habitat, results, and discussion. Instead, descriptive accounts were more literary in organization with beginnings, middles, and endings. However, early ornithologists made extensive use of headings, particularly in articles focusing on specimen studies and bird lists, as seen in Figures 5.2 and 5.3. Otherwise, articles consisted of full pages of descriptive or narrative text.

Graphics

Graphical display was more varied than formatting in early American ornithological discourse. Photography was used to show nests and habitats, whereas drawings were used to contrast physical features of birds, as, for instance, feather and bill shapes. Most striking was an attention-getting color drawing of a Black Rail in the January 1900 issue. Because of early ornithologists' emphasis upon bird lists, lists were the most common method of presenting data. There were also more formal presentations of data using modern tables. However, reminiscent of summaries in early European ornithological discourse, William Palmer (1900, p. 217) displayed subspecies groupings of Maryland yellowthroats by means of a bracketed outline like those used during the Renaissance by Peter Ramus (Tebeaux, 1991a; Tebeaux & Killingsworth, 1992, pp. 19–21).

1900–1940: The Predominance of Natural History Discourse

As the time lines in chapter 4 show, the emergence of natural history study in early twentieth-century American ornithology is striking. This surge of interest in natural history, and the resulting discourse, is at least in part a reflection of the decreasing activity in the specimen laboratory during the period. Because American ornithologists had been successful in describing and classifying the birds of the United States, ornithologists were able to return to the field to study, and to write about, the living bird, as seen in *The Auk*'s 1910, 1920, and 1930 volumes.

1910

By 1910, the pendulum had swung decidedly in favor of natural history, since, of the 35 articles published this year, only two were technical ones identifying two new species and subspecies of birds from lab specimens. Ornithologists' writing was very similar in form to the preceding period, continuing to focus on personal narrative accounts. For instance, natural history accounts predominated with additional personal annotated bird lists, including those from Chaffee County, Colorado; Ashland, Virginia; and Saskatchewan, Canada; from Illinois of its breeding birds; and from Wayne County, Michigan, of its winter birds. There were also personal records of the nesting of Palm-leaf Orioles and Brewster's and Black-throated Green Warblers, of eider courtship, and of Canada Goose migration.

Indicative of a professional orientation, discussions of methodology appeared in an explanation of the newly implemented method of bird banding as an aid in the identification of bird migration. But much of the writing remained strictly narrative, as this example from "Bird Photographing in the Carolinas," which has more to do with the author's personal experiences than with either ornithology or photography:

> The next day, June 11, was spent at the Secessionville rookery in a largely fruitless effort to secure photographs. During a considerable portion of this time Mr. Abbott remained on Morris Island, securing some very good photographs of Least Terns about their eggs. We were caught on the rookery island by low tide and with great difficulty waded the "soap flats" to the row boat [sic] that was to take us back to the "Ethel", being obliged to leave part of our outfit and return for it about one o'clock a.m. (Bowdish & Philipp, 1910, p. 306)

Shown also was a continued interest in ornithological history with the publishing of correspondence concerning Audubon's trip to study the birds of Labrador and a summary of three centuries of records on the rare Passenger Pigeon (a species to be considered extinct in 1914). Finally, there was again another polemic, this time by ornithologist William Brewster (1910), one of the three organizers of the American Ornithologists' Union, in defense of a taxonomic classification that he had proposed:

> . . . he [ornithologist Dwight] pays me a graceful if qualified compliment by asking: "If an ornithologist of Mr. Brewster's ability can go astray in his conclusions what may not the rest of us do?" . . . Thus, like some poor crow, shot and hung up in a cornfield to keep others of his wanton tribe from molesting the precious grain, am I singled out and conspicuously branded to serve as a wholesome example to the ever increasing horde of reckless describers. (p. 324)

1920

In the January issue of this volume, Editor Witmer Stone (1920) provided an instructive description of the process by which articles were chosen for the journal.

> At the outset of a new year "The Auk" finds itself with an abundance of material on hand . . . [so] the appearance of some of the papers will of necessity be delayed. . . . and as [*The Auk*] appeals to a very wide range of readers, it is necessary to keep the matter in each issue as varied as possible. What might be called "readable" articles are therefore arranged in one series and technical papers and geographic lists in another and the aim of the Editor is to mingle the two judiciously in every issue. *If one predominates it is evidence that material of the other kind is lacking* [italics added]. Moreover in accepting papers a wide range of qualifications is considered, for it seems that everything that pertains to ornithology should have a place on the pages of "The Auk" if it is to be, what we hope it may become, the leading ornithological journal in the world. Therefore papers are accepted for their historic, literary, biographic and economic value as well as for their intrinsic scientific worth. (p. 182)

Several of Stone's comments support the findings of this study. First, there was a clear intent to intermingle both amateur and professional discourse in *The Auk* at this time. Second, bird lists were considered part of technical ornithology. Finally, from the italicized sentence above, the preference for natural history over technical ornithology during this period can be inferred. That is, since the journal was heavily oriented toward natural history discourse during this second period of American ornithology, as shown in chapter 4, then the journal must not have received enough "technical" papers to balance the material offered in the journal.

In this same notice, Stone (1920) announced that, at the request of readers, *The Auk* would now publish addresses of authors at the end of articles. Thus, it may now be possible to determine the percentage of amateur versus professional contributors to *The Auk*. In this volume 44%, or 14 of 32 authors, listed home addresses; thus, it may be supposed that almost half of the contributions were from amateurs (that is, those not affiliated with a university, museum, government, or zoological park). It is recognized, however, that such percentages may only approximate professional affiliation. A home address does not exclude the possibility that an author also has professional affiliation or that the quality of such an author's work was not of professional quality.

Again reflecting this large percentage of amateurs, the volume contains the persistent personal accounts of natural history, as for instance 10 articles on the occurrence of birds in a given location and three discussions of the habits of specific species of birds. Articles still abounded with the traditional narrative structure, as seen in this conclusion to an account of the birds of the Catskill Mountains:

> The dusk deepens, and the chorus increases till all the shadowy forest is echoing with deliciously clear music. Then, as darkness falls, they hush one by one; the sky fades over the western mountain; a Great Blue Heron flaps heavily up the lake and over the now silent forest, and far up the valley the "Whoo-hoo-hoo-ah!" of the Barred Owl floats down to us, mellowed by distance, telling that night has come. (Cobb, 1920, p. 49)

However, articles discussing occult senses in birds, courtship in birds, and bird song in shorebirds, though still personal, speculative ones, imply that interest was beginning to pass from strictly empirical identifications of the birds of an author's immediate environment to broader, theoretical schemas concerning the ethology of birds. That is, the author chose to write on courtship *in birds*, rather than the courtship *of a particular species of bird*, as seen in this account of sequestration in birds, which generalized based upon individual observations:

> As it thus forages, each Kinglet every now and then utters its note, or series of notes. Another individual, or others, may be heard from time to time in the distance, but I have failed altogether to receive the impression that two or more birds "answer one another." My experience is that they most certainly do *not* come *towards* one another as the result of such calls. And here the idea presents itself, logically, that these notes serve to keep the foraging birds *apart*: they are *sequestration* notes. (Grinnell, 1920, p. 86)

Still, the 1920 *Auk* allowed its pages to be used for personal polemic, such as this scolding of a colleague by H. S. Swarth (1920):

> I am, however, perfectly justified in feeling resentful at the ungracious wording of his [J. D. Figgins'] argument. I object to such statements, for

> example, as that measurements I have taken are unreliable and that I have suppressed such measurements as did not answer my purpose. I object to having statements ascribed to me that I did not make. I object to having statements of mine "interpreted"—I do not think they need it. (p. 268)

On the technical side, eight articles focusing on the study of specimen skins appeared in 1920, including a proposed new race of Killdeer and a lengthy 30-page article on revising a genus of quail, the article's length representing the continued emphasis upon systematic ornithology. Ornithologist-historians consider the 1920s a defining period for ornithology, because zoologists began to realize both the interconnection among disciplines and the value of birds in the search for causation (Sibley, 1955, p. 630; Stresemann, 1951/1975, p. 351). Chapter 6 will further examine the discourse during this decade in order to evaluate the importance of the 1920s for American ornithology.

1930

In the 1930 *Auk,* a slightly smaller percentage of authors (38%) listed themselves as unaffiliated. Still, natural history accounts predominated, but dropped off in number, accounting for exactly half (20 of 40 articles) of the volume's content. There were the usual array of personal accounts of nesting studies and bird migration, including an account of a Texas birding trip, a topic relegated today to amateur birding publications. The persistence of amateur discourse in *The Auk* is perhaps best illustrated by a study of the American Dipper by Clyde Ehinger (1930), whose motive for study was neither professional nor scientific, but personal:

> For years one of the writer's most cherished ambitions in the avian field, has been to obtain a practical first-hand acquaintance with that unique water sprite, the Dipper or Water Ouzel. This consuming desire was chiefly excited by reading in John Muir's work "The Mountains of California," the chapter on the Water Ouzel. (p. 487)

Also still present were the traditional articles describing specimen studies, but now restricted to only three entries, reflecting the greater stability of American ornithological nomenclature.

Noticeable in contrast was a much greater diversity of topics. There was a movement to the laboratory with a University of Illinois analysis of regurgitated owl pellets to determine dietary needs. A superficially typical annotated bird list of the birds of St. Martin and St. Eustatius Islands was permeated with references to stomach contents of listed species, the author not satisfied with the traditional account of the more subjective, natural-history-oriented occurrence and behavior notes: "One stomach contained mostly grasshoppers; another a large wasp, a fleabeetle (*Homophaeta equinoctialis*), and some other insect fragments" (Danforth, 1930, p. 46).

Instrumentation, in addition to the binocular and ruler, was now part of the ornithologist's tools, a crucial development in the professionalization of a science and its discourse (Latour, 1987). For instance, one writer employed a clinical thermometer to obtain body temperatures of nestlings and another described an "itograph," a mechanical instrument devised by the authors to monitor nesting activities whereby the bird trips a switch as it enters its nest, thus initiating a written record of the event. The authors' simple, straightforward example of the process was accompanied by five photographs of the invention (see Figure 5.4):

> . . . when the adult bird comes to the nest in the box, it must first alight on the outer perch, pass through the entrance, and step on the inner perch before it reaches the nest. When it alights on the outer perch, its weight is sufficient to press down the perch so that it rests on the support below. This establishes a contact between the two, thereby closing an electric circuit. . . . (Kendeigh & Baldwin, 1930, p. 474)

Economic ornithology appeared in a study of birds as a factor in the control of stomach worms in swine. Two historical articles show continued interest in the past, with one discussing a Linnean description and another a collection of eighteenth-century paintings of Haitian birds, as well as an article on paleornithology discussing the age of Cretaceous birds. Ecological awareness was the focus of a University of California at Berkeley study of hummingbird color preference, in which the author demonstrated the interrelationship between floral and bird biology. A foreshadowing of the concern later in the century for human disturbance to the environment appeared as authors discussed both the detrimental effect of artificial structures such as lighthouses on migrating birds and the effects of oil pollution on birdlife, the latter reading more like an investigative report in *Time* or *Newsweek* magazine:

> Oil pollution, however, is not confined to coastal waters, and it may be safely stated that all agencies engaged in the production, transportation, handling, or use of oil, must be considered as actual or potential sources of pollution. . . .
>
> The large earthen tanks or reservoirs sometimes used in the oil fields also may become the means of destruction of hundreds if not thousands of birds. . . .
>
> What is the answer? Many States have passed stringent antipollution laws which if vigorously enforced would do much to relieve the deplorable conditions existing in vast areas of coast, river, and marsh. (Lincoln, 1930, pp. 548–549)

Also foreshadowed in this volume is the increasing empiricism in ornithological discourse, as seen by this excerpt from a study of Black-throated Blue Warblers by Margaret Morse Nice (1930):

FIGURE 5.4. Example of instrumentation in early American ornithology (Kendeigh & Baldwin, 1930, p. 474). Used with the permission of the American Ornithologists' Union.

> Broodings varied from 2 to 20 minutes in length (except for one of 41 minutes on the afternoon of June 26 during and after a rain); the average time was 9.1 minutes. The percentage of time spent in brooding was 22.2 on June 25 and 18.9 June 26. She brooded after half the feedings on June 25 and after 22 per cent of them the next day. (p. 341)

The 1930 volume was also more internationally oriented, including a study of the breeding birds of China, a life history of a South African species, and an article by a Russian ornithologist discussing a new race of Russian bunting. Finally, there were attempts to standardize the study of ornithology. For instance, one article proposed a method to systematize oological data, whereas another was essentially a laboratory recipe for preparing study-skin specimens. Most significant were two early references to statistical methodology, which would eventually permeate ornithology: One article proposed a methodology for recording bird counts for statistical purposes and another discussed the use of statistics in the study of bird song, which included an early example of the use of line graphs in *The Auk* (see Figure 5.5). Thus, the increasing diversity in American ornithology from 1920 to 1930 is strikingly evident.

Organization and formatting

During the second phase in the development of ornithological discourse between 1900 and 1940, a structured Introduction, Methods, Results, and Discussion (IMRaD) format was beginning to appear, though it was far from a standard format. The articles in the 1910 volume had the same general appearance as those in the volumes from the first period. That is, there were generally full pages of text except for bird lists. Seeing the need for a guiding structure, some authors did subdivide their narratives into meaningful section headings like *Nest Configuration*, *Feeding*, and *Singing*. One author, in fact, used the word *Results* as part of his heading: *Results of the Season's Work* (Cole, 1910, p. 158).

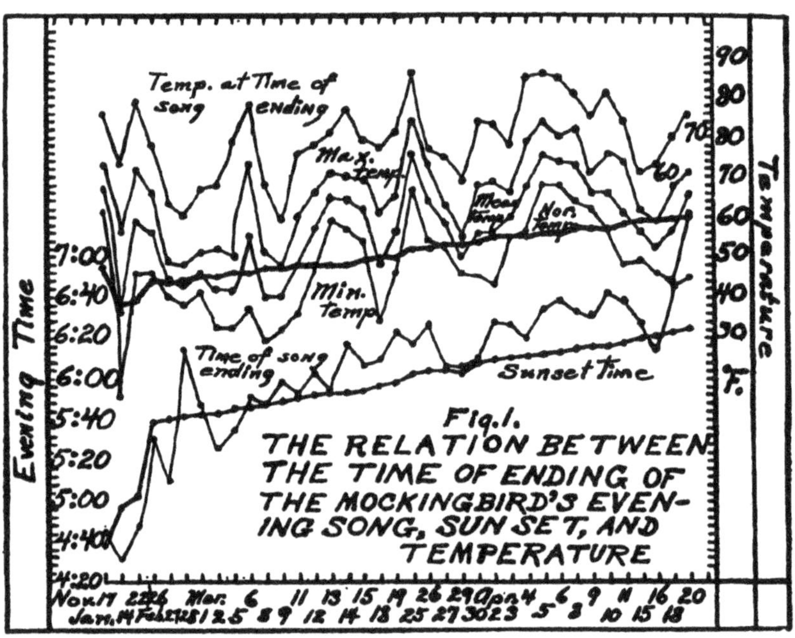

FIGURE 5.5. Early line graph in American ornithological discourse (Shaver & Walker, 1930, p. 388). Used with the permission of the American Ornithologists' Union.

The 1920 volume contained the first evidence of the use of standardized formatting, which was employed by three authors, two of these professionally affiliated. J. D. Figgins (1920) of the Colorado Museum of Natural History was the first in this survey of 10-year volumes to use a standard heading (*Conclusions* on page 101). In the next issue, Alexander Wetmore (1920) of the U. S. Biological Survey used two: *Introduction* (p. 221) and *General Discussion* (p. 227). C. K. Averill's page and a half *Summary* (1920, pp. 578–579) was really a list of conclusions from his data.

In 1930, only five authors used one standard heading each, though two of the headings (*Literature Cited* [Harper, 1930, p. 185] and *Methods* [Gardner, 1930, p. 368]) are new to this survey. That four of those using headings gave professional addresses may indicate an academic origin for this preliminary standardization of ornithological discourse. In this volume also, Eloise Cram (1930) of the U. S. Department of Agriculture was the first author surveyed to structure her entire article in a standardized format by using *Introduction* (p. 380); *Findings of present investigation* (p. 381); *Discussion* (p. 383); *Summary* (p. 384), a true summary; and *Acknowledgments* (p. 384). Though perhaps superficial, this preliminary evidence that ornithological discourse was becoming scientifically structured reflects a changing discourse, as will become more evident in the discussion of the third period in the development of ornithological discourse.

In this early phase of American ornithology, the journal did not yet provide specific instructions or suggestions for contributors to *The Auk*. In fact, from the description by Witmer Stone (1920) of the journal's editorial procedure, it appears that most all manuscripts were automatically accepted: "It seems that everything that pertains to ornithology should have a place on the pages of 'The Auk'" (p. 182). The same is true in 1930, judging from the note on the inside back cover of the journal that article manuscripts must merely "await their turn for publication if others are already on file" (*Auk*, 1930). In 1920, Stone even suggested establishing a $25,000 endowment for the journal so that it might double in size in order to accommodate the growing number of manuscript submissions. But the size of the journal did not change significantly during this period.

Graphics

Though the discourse of the period was dominated by full pages of text, graphics continued to be popular additions to American ornithologists' writing. During the period 1900–1940, listing remained relatively popular due to the continued use of annotated lists of birds, whereas the number of tables decreased slightly. Drawings registered the largest increase over the first period of American ornithological discourse, particularly in the 1920 volume, which included 23 graphics containing drawings, like those by Allan Brooks of Golden-Eyed Ducks (see Figure 5.6). Significant as predictors of future graphical displays in ornithological discourse were two line-graphs appearing during this phase, one of which is shown above in Figure 5.5. The use of photographs also registered a large rise in popularity among ornithologists, who used photographs in a total of 58 figures in the 1910, 1920, and 1930 volumes. Characteristic of the emphasis upon natural history during this period, as well as the limitations of the camera itself, habitat and nest sites were the most commonly chosen subjects for photographic display.

1940–1970: An Interplay of Discourse

The rapid advances in science and technology during the post-World War II era brought corresponding sophistication to American ornithology, as the 1940, 1950, and 1960 volumes demonstrate. In the 1950 volume is also evidence that *The Auk* was becoming more restrictive with regard to the articles it published: "For publication in 'The Auk,' articles or notes must present material that has not been published elsewhere and that is of significance to ornithologists" (*Auk*, 1950). Still, with the influence of observation-based

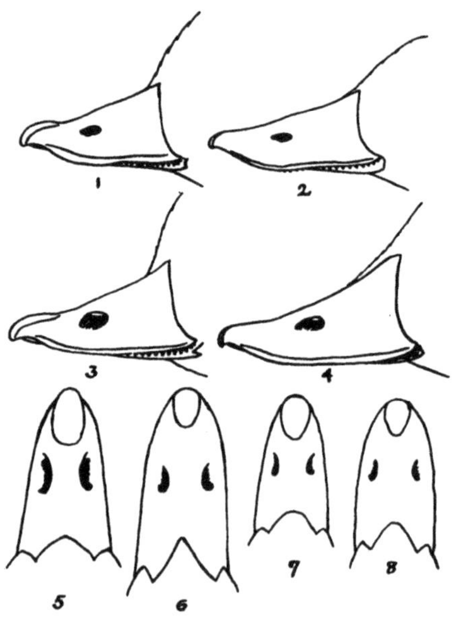

FIGURE 5.6. Example of a series of drawings, a popular graphic during American ornithology's natural history period (Brooks, 1920, p. 360). Used with the permission of the American Ornithologists' Union.

study, natural history writing continued to play a prominent role in American ornithology.

1940

As sophistication increased, natural history discourse correspondingly decreased in the 1940 *Auk*. But, even though only about one-third of the articles published during the year were devoted to the natural history of birds, the importance of this discourse is still evident. For example, "Summer Life of the Sora Rail" headed issue 2, "Social and Sleeping Habits of Central American Wrens" began issue 3, and an article detailing the social nesting habits of Smooth-billed Anis was given 39 pages of text, far more than the average article. The description by Lawrence Walkinshaw (1940) of his observations of Sora Rails sounds very much like the natural history writing of the first phase of American ornithology:

> At one nest, on May 27, 1934, where I was attempting to photograph the adults, I had captured and placed in my hat within the blind, five downy young, hoping that the parent might return quicker to the eggs. After waiting two hours without success, although both parents were walking through the sedges very close to the blind and the young were occasionally peeping from inside, I released them, one at a time, with the result that one parent assembled them a short distance from me while the other returned to the edge of the nest, but would not incubate the eggs while I was there; so I finally left. (p. 166)

Still acceptable were accounts of bird trips, annotated lists, and personal accounts of birds seen during the severe winter of 1939–1940. However, demonstrating the period's interplay of discourse, almost half of the articles published in 1940 described ornithological work conducted in the laboratory. Though taxonomic studies of preserved bird skins still dominated this phase of ornithology, as shown by the 10 of 14 taxonomic articles constituting the 1940 volume, the lab was now seen as a place to study the living bird. Described, for example, were studies of avian infections, as well as of caged weaver finches to determine stages of molt in birds. The writing in these articles appears directed toward a technically oriented and empirically inclined audience, as in the excerpt from an article describing the factors affecting incubation length:

> If the values for all temperatures are averaged together for each age, the following variations are noted: 0 days, 3.4; 2 days, 0.78; 4 days, 0.64; 7 days, 0.69; 10 days, 0.72; 12 days, 0.72. There is considerable difficulty in using the respiratory quotient as a criterion of the type of food being metabolized, although for the two older ages, one may suppose that fats are involved. The very high values for unincubated eggs suggest that carbon dioxide is

eliminated in excess of that freed by energy exchanges. . . . (Kendeigh, 1940, p. 504)

In addition, field ornithology had now become more complex. This is the first volume in this survey in which articles describing detailed field measurement are numerically equal with simple natural history accounts (6 articles of each). Ornithologists were now manipulating birds experimentally, as shown by the inclusion of three articles describing this new methodology. In one homing study, petrels were captured and later released at various distances from their nest sites. Another study described an attempt to manage birds by introducing nest boxes. In the most elaborate of the three studies, eggs of Laughing Gulls were manipulated as part of 32 separate experiments to determine the gulls' egg recognition, as explained in this excerpt and shown in Figure 5.7:

> Text-figure 1 shows the reactions of the birds to visual cues, namely, the birds' first reactions, based on what they see. The subsequent behavior may be affected by tactile cues. In Experiment 8, where a bird sat on a spot where there were no eggs, the birds were seen to become steadily more uneasy; finally the birds, when the eggs were 12 inches from the nest, rolled them in; when they were 32 inches away, the bird, after brooding the empty nest twenty minutes, went to the eggs. (Noble & Lehrman, 1940, p. 40)

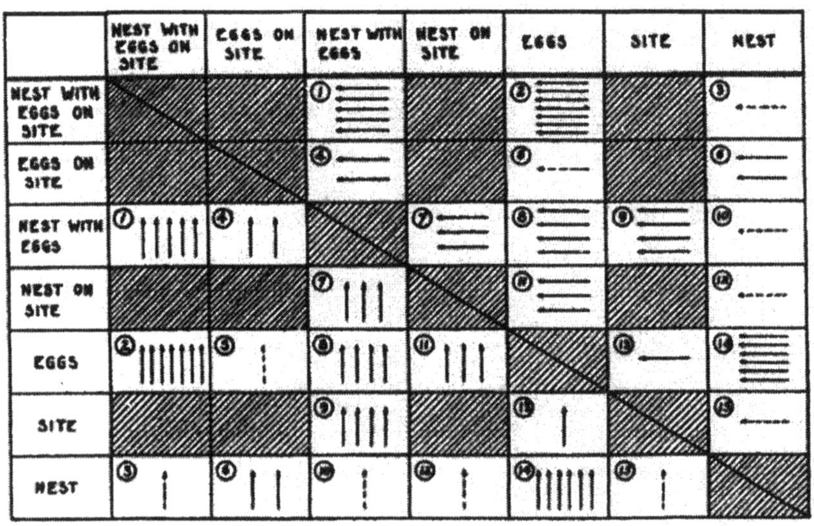

FIGURE 5.7. Response matrix demonstrating an elaborate ornithological study conducted in the field (Noble & Lehrman, 1940, p. 24). Used with the permission of the American Ornithologists' Union.

Despite this added sophistication, articles of a theoretical nature were still lacking. Only two appeared in 1940—one discussing a theory of courtship feeding and a second presenting a mathematical formula for deriving the relationship between egg volume and incubation period:

> . . . a mathematical analysis . . . yields the following relationship between egg volumes and incubation periods:
>
> $$V = .0063\ (1000\ V / T\ )^{1.30}$$
>
> Let us test this formula by calculating the probable incubation period of any egg simply from its dimensions. The following case is an example, selected at random from the list of birds which fall on the line of Text-fig. 2, and for which, therefore, no ecological correction need be made. (Worth, 1940, pp. 51–52)

1950

The 1950 volume continued to reflect the interplay between natural history and experimental science discourse. There were the persistent taxonomic discussions over ornithological nomenclature, including five articles describing specimen study-skins in detail. In addition, studies of growth bars in feathers appeared, as well as an attempt at behavior modification of hand-raised crows. On the theoretical end, there were presentations of a migration theory and an attempt to use a logistic curve to describe mathematically the population increase of starlings:

> These data, smoothed by moving averages of threes, are plotted on semilogarithm paper by years for New York, Massachusetts, and Ohio (Fig. 2). It is at once apparent that the populations increase logistically to an asymptote of about 40 Starlings per hour for New York, 35 for Massachusetts, and perhaps 20 for Ohio. . . . it should be mentioned parenthetically that a more elaborate treatment of the theory of the logistic predicts inherent oscillations about the asymptote. (Davis, 1950, p. 462)

During a status review of the concept of "species," an author argued for ornithologists' role in the clarification of the term—a veiled attempt to assert ornithology's professionalization. However, a substantial number of articles were still primarily personal discourse: annotated bird lists, bird records, nesting records, field observations—thirteen articles (48%) devoted to topics in the realm of the natural history of birds, though empirical measurements predominated in many of these accounts. Particularly telling is the introduction to "The Sandhill Crane in the Bernard W. Baker Sanctuary, Michigan," an account of personal visits by the writer, who gave the impression that he was merely presenting data left over from a previous publication:

> In September, 1921, I made my first visit to the area now constituting the Bernard W. Baker Sanctuary. Several visits were made to the area during August, 1930, and several during May, 1931. Thereafter, I have visited it nearly every month, often many times during some months. Some of these observations have been published by the author (1933, 1941, 1944, 1945, 1946, 1948). A great deal of information on the Greater Sandhill Crane, *Grus canadensis tabida* (Peters), has been secured on the area. Some of this material has been incorporated in my manuscript on the Sandhill Crane, but a great deal of it could not be used. I am assembling that data in this article. (Walkinshaw, 1950, p. 38)

Three historical articles indicated that the field was still concerned with its ornithological past; in other words, American ornithology was still searching for its identity. The most revealing indication of continued amateur influence was the publication of a speculative account of the flight mechanism of swifts and hummingbirds in which the author used a homemade stroboscope, as he described:

> A simple, clockwork, rotating shutter was arranged to cover the right objective of a pair of binoculars. By watching birds in direct flight toward or away from me and by closing the left eye, it was often possible with some practice to "stop" the wings for two or three beats, which was ample to show that they moved in unison. Actually, after about two months of almost daily observation, it was found that an increased ability to analyze motion allowed all but the most complex evolutions to be followed without the stroboscope. (Savile, 1950, p. 499)

Thus, in this one journal volume, the range of expertise is striking: from purely personal observations of bird migration to an extreme case of field measurement by quantifying eye diameter and length of individual toes and feathers, from accounts of ornithology's history to an assertion of professionalization, and from analyses using both a homemade stroboscope and mathematical logistic curves. Unquestionably, the break between natural history and experimental science discourse had not yet occurred.

1960

In contrast, the 1960 *Auk* had become dominated by professional discourse, if the number of professionally affiliated contributors was any gauge of the professionalization of American ornithology. In 1940, three-fourths of its authors gave professional addresses, whereas by 1960 89%, or all but three authors, were associated with a government or academic organization. This decided change in the contents of the journal may, at least in part, be a function of a decision in 1954 by *The Auk* to judge manuscripts "primarily for their future value and secondarily for their present interest," which, according to *Auk* editor Robert W. Storer, should result in an emphasis on the habits and biology of birds, rather than on their occurrence (Wolfson, 1954, p. 75). Of course, occurrence records have traditionally been a principal contribution of the amateur.

Reflecting this increase and the approaching end of the period of discourse interplay, articles based upon laboratory research were the most numerous in 1960 (43%). There were studies to determine jaw articulation, fat-free weights, energy requirements of Tree Sparrows, egg temperatures of Ring-necked Pheasants, the functioning of Wild Turkey gizzards, and syllable structure in the song of the Brown Towhee. In addition, a Harvard University researcher described an expansive study of the avian medial brace, a bony support of a bird's mandible. Here is an excerpt of the study, which comprehensively analyzed skulls from 85 families of birds:

> Spheniscidae. (a) The mb. is present in several genera of penguins, such as *Aptenodytes*, with the mpm. articulating against the mbp., but it is absent (?) in others such as *Spheniscus*. Perhaps the articular pads in these latter genera are thick and have dried out, which would account for the apparent absence of the mb. (b) The lbp. is present as a poorly defined ridge along the margin of the basitemporal plate. (Bock, 1960, p. 40)

Most notable is the use of sophisticated equipment to provide solutions to proposed problems. For instance, University of Georgia researchers used a vacuum oven to compile a table of fat-free weights in birds for use by bird banders as an aid in estimating fat levels in the field. A Canadian researcher studied captive Tree Sparrows' energy requirements using a "Parr adiabatic oxygen bomb calorimeter . . . with a Weston Illumination Meter . . . with viscor filter" (West, 1960, pp. 308–309). Multiple measurements for recording the temperature of Ring-necked Pheasant eggs at 16 different locations every 30 seconds was achieved by an Ohio State researcher using a "Leeds and Northrup Model G-S-6000 automatic recording potentiometer . . . [with] 24-gauge copper constantan thermocouples with extruded nylon insulation" (Kessler, 1960, p. 331). Needless to say, these were technologically far advanced from the 1950 account of the use of a homemade stroboscope.

As is evident from this abbreviated list, the pendulum had swung, at least in laboratory ornithology, from taxonomy and observation to avian physiology. Only two of the 10 laboratory articles were descriptions of newly collected specimens from fieldwork. Erwin Stresemann (1951/1975) asserted that such interchange among scientific fields, such as between ornithology and physiology, precipitated ornithology's rapid advancement during the twentieth century (pp. 351–364). If this assertion is correct, then perhaps this volume indicates the beginnings of the emergence of

American ornithology as a professionalized science, a question to be addressed further in chapter 6.

But still in this same volume, juxtaposed with these technical subjects, was an article on a birding vacation to Mexico by the author and his wife. The article stands in sharp contrast to the eight laboratory research articles published in the same volume. It begins,

> Of the 10 birding vacations that Mrs. Coffey and I have undertaken in Mexico since 1946, five have been in late spring and one in late April. As a result, one of our special interests was that of observing the extent of the spring migration of North American species through eastern and southern Mexico. . . .
>
> From 20 May through 10 June 1951, we went as far as the highway's end just south of Catemaco, Veracruz. From 17 May through 6 June 1953, we rushed to the end of the Inter-American Highway, at the Guatemalan border, then lingered at Las Casas, Chiapas, and detoured to Veracruz. (Coffey, 1960, p. 288)

In the next issue, a methodology not ordinarily associated with contemporary empirical science was described. In order to study the spread of the Cattle Egret in the United States, questionnaires were mailed to bird clubs and letters sent to "special persons . . . and authors of note," in addition to checking *Audubon Field Notes* (Davis, 1960, pp. 421–422). This author's reliance on amateur observations by the use of information from bird clubs and the Audubon Society, as well as another article reviewing Passenger Pigeon records from Missouri, indicated the continued use of casual bird observations in 1960 ornithological science.

The presence of natural history discourse was likewise evident in observation-based articles in the volume, such as nocturnal migration of thrushes; foraging, courtship, and territorial behavior; and a breeding study. Five of the nine, however, do have an empirical orientation since they reference counts and/or measurements.

Theoretical issues were still absent, except for a theory of coastal migration and a review of the literature on weather-related influences on migration. The latter article, posing theoretical questions and suggesting answers, included an impressive 151-entry bibliography.

Organization and formatting

During this third period in the development of American ornithological discourse, ornithologists still had no standardized format, though it may be argued that the bird lists and specimen studies written during this period still followed the organization used in the earliest volumes of *The Auk*, as shown above in Figures 5.2 and 5.3, and were consequently the "standard format" of American ornithology. Still, the 1940 and 1950 volumes do not demonstrate any organized use of the traditional standard format, though the standard headings (Introduction, Methods, Results, Discussion, Conclusions, Summary, Acknowledgments, Literature Cited) were beginning to filter into American ornithologists' writings. In 1940, for instance, 53% of authors and in 1950 77% used at least one standard heading. In lieu of this, many authors used more informative headings to subdivide their articles. It is also interesting to note the increasing use of Acknowledgment sections, which signal an awareness by ornithologists of collegiality, of a sense of a professional self:

Year	*Occurrence of Acknowledgments*
1940	1 (3%)
1950	4 (15%)
1960	10 (44%)

The year 1960 also shows a move toward professionalization, and a professional discourse, by the almost universal use by ornithologists of at least one standard heading—all but one of 23 articles. In fact, the format used by George C. West (1960) in the October issue (unheaded Introduction/Literature Review, then headed Methods, Results, Discussion, Acknowledgments, Conclusion, and Literature Cited) is very similar to what would become the standard format in *The Auk*.

Significant changes also occurred during this transition period with regard to the journal's stylistic prescriptions. In 1940, direction was now given regarding typewritten manuscripts, the handling of citations, submission requirements for drawings, admonition against the use of "extensive" tables, and the need for "longer articles" to end with a brief summary (*Auk*, 1940). In 1950, formatting information became more detailed regarding spacing, margins, titles, photograph and table submission, and naming of birds. In an important move toward standardization of formatting, the October 1961 *Auk* announced that the journal would join 78 other biological journals by following "in great part" the *Style Manual for Biological Journals* newly published by the Council of Biology Editors (Notes and News, 1961, p. 669). Beginning in 1963, the Suggestions to Contributors, now a full page, gave a detailed description of the format for manuscript submission, covering such topics as line spacing, type of paper, margin limits, quality of graphics, wording of titles, reference section, and formats for citing birds' names, weights, and measurements (Suggestions, 1963).

Graphics

During the period of discourse interplay (1940–1970), there was an 18% increase in the use of graphics. Lists still remained at about the same level of popularity, showing the persistence of bird lists and specimen studies, which traditionally employed listings. Drawings were still used with about the same frequency but in more diverse ways to demonstrate such activities as bird behavior and wing motion. Though maps were on the increase, the use of photographs began to decline, becoming almost nonexistent by the end of the period. The 1960 volume had only three, whereas previously there were on average 15 photographs per volume.

The most significant evidence of a shifting discourse is the rapid rise of two other types of graphics—tables and graphs. The use of tables, both formal and informal, registered a six-fold increase from an average of seven per volume during the previous two periods to 53 per volume during this third period of American ornithological discourse. The greatest increase, however, was in the use of graphs, particularly line graphs. Whereas only four graphs were used in both previous periods combined, writers of articles published in the 1940, 1950, and 1960 volumes chose graphs 39 times to display their data. There was likewise a diversity of graphs, from line, bar, and area graphs to scatterplots and hybrids. The line graph in the October 1960 issue illustrated in Figure 5.8 foreshadowed the rapid rise of statistics in American ornithology, because it is the first example in the 10-year volumes surveyed the sole purpose of which was to illustrate regression lines,

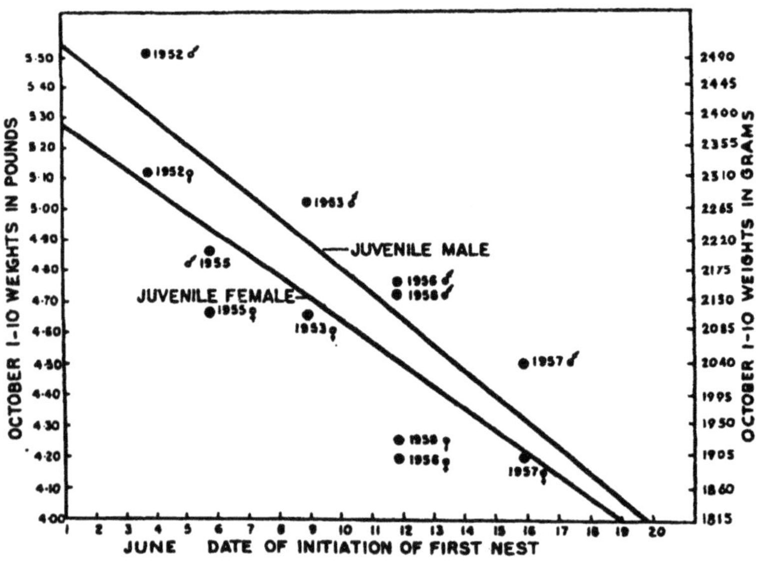

FIGURE 5.8. Example of a line graph, a prominent graphic during American ornithology's third period of discourse (Cooch, Stirrett, & Boyer, 1960, p. 462). Used with the permission of the American Ornithologists' Union.

statistical predictors that will overtake experimental science discourse during American ornithology's fourth phase, to be described next.

1970–1990: The Emergence of Experimental Science Discourse

Just as the computer has forever changed life in general, it has altered the face of contemporary American ornithology and its discourse, by providing the calculating power to make inferential statistics a common methodology in American ornithological science. This rapid rise in statistical argumentation after the 1960s was a significant factor in the increasing insularity of American ornithology. Richard Whitley (1977) has described the impact of arithmetic reasoning and mathematical training on the professionalization of the physical sciences, and by extension science in general, whereby such expertise distances these fields from untrained, lay audiences. Similarly, it was the widening use of statistical analyses that ultimately provided the impetus for the emergence of American ornithological writing from primarily natural history to experimental science discourse. It should also be noted here that the journal was now becoming increasingly selective in the manuscripts it accepted for publication. In 1967 and 1968, *Auk* editor Robert Mengel noted, for instance, that the journal's rejection rate was at 50%, the first such reference to *The Auk*'s increasing exclusivity (Mewaldt, 1968).

1970

In 1970, the dichotomy between natural history and experimental science discourse remained; in fact, there appears to be a resurgence of natural history discourse. For instance, the 1960 volume contained only 9 observation-based articles out of 23, whereas in 1970 18 of 42 articles focused on natural history study. Most striking was an article by an apparently unaffiliated author who used her own backyard to study anting by birds, supplementing her analysis with meteorological data from a local airport:

> From the glass-paned back door of my home in Zebulon, I could scan effectively without aid of a binocular a section of lawn approximately 75 feet square. I tried to note the behavior of all birds within this area at least once every 2 hours from 06:00 until dark almost daily from 3 March 1964 through 31 August 1966. (Potter, 1970, pp. 692–693)

Despite its methodological simplicity, the article contained elaborate empirical documentation after the manner of the ground-breaking behavioral work done also at home three decades earlier by Margaret Morse Nice, whom the author referenced.

Also in the 1970 volume was a form of writing indicative of natural history articles published during the early years of *The Auk*. At that time, there appeared many articles based on personal observations (see Figure 5.2), often entitled "Notes on" some geographical location or species. In 1970, there were two more "Notes" ("Notes on the Seabirds of Round Island, Mauritius" and "Notes on the Red-cockaded Woodpecker in Texas"), giving the journal an air of the past—that is, back to descriptive ornithology. Perhaps such natural history presentations were acceptable because of the novelty of the study area in one case or the endangered status of the species in the other. Yet three other articles also took the reader back in *Auk* history. "Mating Behavior of the Common Loon" reads like an article from decades past, giving personal narratives in first and third person, as does this excerpt from "Bathing Behavior of Land Birds":

> For several years I noted that the Flickers (*Colaptes auratus*) on the old Long Island farmstead bathe in the mudponds by using partly submerged logs and sticks as supports. In contrast, the Flickers in the Oxon Run tract and Cedarville Forest bathe in rain. Only once did I see a Flicker bathe in a shallow stream, and its actions were clumsy. The body rolling was slow, superficial, and somewhat out of phase with the flicking of wings. (Slessers, 1970, p. 96)

Even the more technical "Summer Lipid Levels of Some Subarctic Birds" added a "Natural History" subsection, which reads like early twentieth-century annotated lists. However, the natural history discourse in the volume was balanced with more empirically based observations of breeding and courtship ecology, territoriality, predation, and an exacting study of Western and Semipalmated Sandpipers, whose feeding behavior was quantified by counting the number of seconds spent probing and pecking in two different types of mud:

> As close as possible to the time of low tide, I made a series of counts of the flock in quick succession, noting first how many of each species were in water, then how many were on wet mud, and finally how many were on dry mud. . . .
>
> I also determined the proportion of feeding time spent in probing, as opposed to pecking, by each of the species within each of the three zones, ignoring birds close to the boundaries (Table 2). Probing I define as feeding by jabbing continually with at least the tip of the bill always below the surface, except for brief moments when the bird changed its probing site; pecking is feeding by pecking the surface, each peck being a clearly distinct feeding movement. (Ashmole, 1970, pp. 131–132)

Indicative of the imminent predominance of experimental science discourse, 16 of 42 articles described technology used in the ornithological laboratory. Some were anatomical and physiological studies, as for instance of lipid levels in three species, conservation and dissipation of heat in Turkey Vultures, sparrows' salt intake, skull evolution in ducks, and quail sexual development. Physics had now entered ornithological research through the field of bioacoustics, which had introduced the sonograph to quantify bird song. In 1970, bird vocalizations, analyzed by studying sound spectrograms, were the subject of five studies, an unusually large number of articles in one volume on a single topic. More importantly for the future of experimental science, statistics and the computer had become tools of ornithology. One researcher offered a predictive regression model for avian water-loss values, and another proposed a mathematical model to show adaptation in leg muscles of two species of ducks. Finally, statistical analysis was the basis for a study of color variation in soft parts and down of Royal Tern chicks. Though only three articles incorporated statistical methods in 1970, their use would permeate the field by 1990, allowing American ornithologists to make more theoretical statements than they previously had. In 1970, for instance, the only other article which put its study in a larger perspective was a museum specimen study of feather tracts, suggesting its use in taxonomic classification in general. After almost a century, the attempts to clarify systematics continued.

1980

Although in 1980 the percentage of natural history and laboratory articles remained about the same, advanced laboratory techniques and the 31% increase in theoretical articles showed ornithology's increased sophistication. Perhaps the increase in theoretical articles was also in response to a specific request beginning 2 years earlier for contributions "dealing with theoretical or methodological developments" (*Auk*, 1977, p. 191). In fact, ornithologist Ernst Mayr (1983), in his introduction to the AOU's centennial volume on contemporary ornithological research, noted the increasing emphasis upon theory and model building by ornithologists (p. 2). Contributing to this change was an editorial decision by *The Auk* to "become more selective in the material that is accepted for publication" (Wiens, 1978, p. 219) in order to decrease the time between manuscript acceptance and publication.

Editorial changes during the decade increased the competition even more. Beginning in 1982, review articles were no longer encouraged, but only "considered" (*Auk*, p. 184). A 1985 editorial commented on articles accepted "from areas not traditionally thought of as primarily ornithological" (Brush, p. 903) and defended the "complex" (p. 903) decision-making process as a search for a "balance of content, page limits, perceived reader's interests, and other subtle factors" (p. 903). Beginning in 1988, manuscript submissions required a letter of transmittal, which included statements that the material was original and that the manuscript had not been submitted elsewhere (Brush, 1988, p. 803).

Particularly indicative of the changes now overtaking American professional ornithology was a volume of essays published in 1983 to commemorate the centennial of the American Ornithologists' Union (Brush & Clark, 1983). Intended as a review of contemporary ornithological research, essays covered a wide range of topics, including biochemical and genetics studies, bird song, bird navigation, avian ecology, biogeography, and cooperative breeding strategies. Calling recent changes in ornithology "profound" (Mayr, 1983, pp. 2–5), ornithologist Ernst Mayr in his introduction to the volume highlighted four of the most important changes: a shift from description to causality, a merger of ornithology with other biological disciplines, a recognition of the interrelationship among biological disciplines, and, most importantly for this study, the professionalization of ornithology. Yet in 1983, Mayr and, symbolically, American ornithology still acknowledged the "indispensable" (p. 3) nature of amateur contributions to ornithological research, though the amateur was now relegated primarily to the role of a data gatherer. (See Lankford, 1981, for an account of a similar, though much more rapid, divesting of authority of British amateur astronomers.)

Thus, in 1980 topics for laboratory study were now more diversified, focusing on anatomy, physiology, morphology, bioacoustics, ethology, and systematics. Indicative of this sophistication was a study of avian energy expenditure, using the following method:

> Blood samples were micro-distilled (Wood et al. 1975) to obtain pure water, which was assayed for tritium activity (Beckman LS 230 liquid scintillation counter, toluene-Triton X100-POP scintillation cocktail) and for oxygen-18 content by cyclotron-generated proton activation of ^{18}O to fluorine-18 with subsequent counting of the gamma-emitting ^{18}F in a Packard Gamma-Rotomatic counting system (Wood et al. 1975). (Weathers & Nagy, 1980, p. 862)

Most indicative of this greater sophistication was new technology to analyze an old topic: systematics. Four articles detailed the use of electrophoresis to study protein characteristics in birds, enabling researchers to suggest evolutionary origins and propose new taxonomic relationships. The specialized nature of the procedure is evident in the Abstract of an article on the evolutionary genetics of birds in the January 1980 issue even from the very first sentence—"We have employed phenetic and cladistic approaches to analyze frequencies of electromorphs encoded by 25–27 loci in several species of North American thrushes" (Avise, Patton, & Aquadro, 1980, p. 135)—and later when the authors described their methods as "horizontal starch-gel electrophoresis. . . . about 300 buffer system X protein stain X tissue combinations . . . in a series of test gels" (p. 138). An additional example appeared in the very next article, also a protein study, in which author Robert F. Martin (1980) included a figure of a "six-phosphogluconate dehydrogenase zymogram" (p. 150).

On another level, the computer had enabled ornithologists to propose mathematical models in three articles: formulas to determine energy requirements, models to predict daily energy expenditures, and a model to assess kin selection. In the 1980 volume were more strictly theoretical articles—19—than in the volumes surveyed thus far. Proposed were evolutionary theories and counter theories, a prediction about bird population dynamics, models of avian community structures, and a theoretical discussion of nests as thermoregulators. Strikingly different in its mathematical and theoretical approach was an article by a professor of physics, who described the effect of the coriolis force on bird flight.

Also evident was the effect of inferential statistics on ornithology, which now appeared more willing to make theoretical statements from what was often basically natural history study, already quantified in previous years but now buttressed by statistical validity. In 1980, principal component, cluster, multivariate, discriminate function, and regression analyses were common methodologies generating their own jargon in the ornithological literature. Additionally, there were covariance matrices, eigenvalues, Euclidian and nearest-neighbor distances, and n-dimensional morphological hyperspaces. Now American ornithologists could make assertions about the structure of an entire avian community in a given ecosystem or estimate weight loss in migrants.

But more often than not, authors did not place their studies in larger contexts, nor argue beyond the statistical claims supporting their particular hypotheses. Only four of the 13 statistically based studies in the 1980 volume did so. In other instances, statistics was often just a support for research that could be classified primarily as natural history study, as for instance studies of foraging areas of Lapland Longspurs, perch-site selection by Red-tailed Hawks, and Antarctic penguin display, all of which were primarily observation-based.

Despite the reliance on statistics, there were still many articles in 1980 functioning primarily as natural history discourse—31% to be exact. Thirteen of these made limited use of statistics; but, unlike those articles just mentioned, their primary function was observation and measurement in the field: counts of roost flights, accounts of breeding and feeding behavior, and the like. As already noted, however, natural history study was now more complicated, as seen in one study that used a 15-second metronome beep to assure standardized recording of observations; and another study of Panamanian antwren flock formation that used stepwise multiple regression and an elaborate method to quantify the amount of vegetation present, including the use of a mathematically based vine tangle index, described by the authors as follows:

> . . . we included a measure of overall density and variability of each class of vegetative structure within each quadrat. In addition we recorded the

> diameter and height range for all dense (estimated to be 75% or more closed to light; 100% would be black) vine tangles between 3 and 18 m. The vine tangle index for each quadrat used in the multiple regression was chosen to reflect the roughly conical shape of vine tangles on the study site:
>
> $$\sum_{i=1}^{n=4} \sqrt{(\text{vertical height range X diameter})}$$
>
> Height to the top of the canopy was estimated, and the estimates were calibrated with a 200-mm telephoto lens. (Gradwohl & Greenberg, 1980, p. 388)

Still, there persisted less complex articles. One author summarized observations of an escaped European Mute Swan, whereas another described a simple backyard study of robins' foraging preferences by observing the effects on robin behavior after the author mowed his lawn:

> *Methods.*—The observation area was a suburban backyard in Lancaster, Pennsylvania. . . . Data were collected daily (except when it rained) during periods that averaged 90 min and that occurred at different times of day ranging from 0700 to 1800; care was taken that the observations did not occur more frequently at some times of day relative to others. The lawn was mowed to a height of approximately 6.5 cm three separate times during June–July 1979; at least 15 days occurred between mowings. . . .
>
> During each observation period, two categories of data were collected: (1) *foraging time*, the number of seconds per hour that a robin was seen foraging on the lawn, and (2) *attack rate*, the number of times per foraging minute that a bird either bill-pounced (Heppner 1965) or snapped at apparent prey in the grass. (Eiserer, 1980, pp. 576–577)

There was likewise another "Notes" article, containing annotated field notes of southern Peruvian birds. Significantly, only one article, "Seventeen Bird Species New to Peru from the Pampas de Heath," can clearly be classified as natural history discourse.

1990

In contemporary American ornithology, however, statistics has fully overtaken ornithological research. Of *The Auk*'s 65 articles, 50 are based on statistical inferences to present their results and conclusions. Still, only a handful of articles address theoretical issues, often merely as concluding remarks; for example, a study of woodpecker nesting closes with a social-science sounding "Implications" subsection. Only two articles are strictly theoretical. One of these, which deviates from standard format to include a "Theory" section, proposes a mathematical model to estimate the effect of wind on avian metabolic rates; a second presents a simulation model to evaluate two methods of estimating nesting success. In the more empirically oriented studies, research from 18 laboratory-based studies continues to focus on the same topics as in 1980: physiology (such as determining cold tolerance, periods of stress, dietary needs, and the role of prolactin in parental care) and anatomy (such as studies of digestive organs and flight muscles). Ever-present systematics studies also continue with starch-gel electrophoresis for investigating genetic variability, but now ornithologists have added a new and controversial technology, DNA fingerprinting, a new method for determining taxonomic relationships by recombining species' DNA chains (See Figure 5.9 for a graphic demonstrating this technology). These advanced techniques have all but replaced the traditional morphological studies for determining familial relationships, as shown by the relatively few lab studies (5 of 18) employing analysis of study skins.

In the remaining 35 studies—over half of all statistically based articles—ornithologists are still in the field. Nineteen of these field studies (54%) are nesting studies, because the nest site, in addition to being an important factor in identifying a species' population dynamics, is simply a convenient place to find and observe birds' behavior and the nestling an easily captured subject for measurement. The other popular field topic in the 1990 volume, species' foraging methods, is complicated by the necessity of

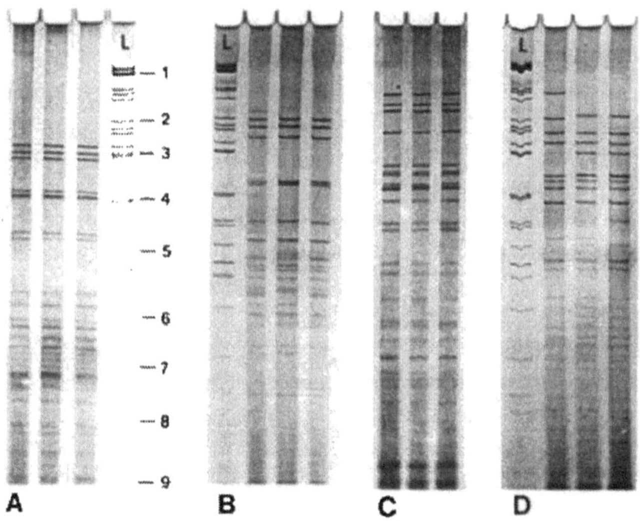

FIGURE 5.9. Graphic illustrating modern DNA research in ornithology (Tegelström, Gelter, & Jaarola, 1990, p. 732). Used with the permission of the American Ornithologists' Union.

following mobile individuals over sometimes prohibitive terrain, which several studies of mourning doves, screech owls, and ospreys overcome by the use of expensive radio-telemetry equipment.

A closer look at 41 observation-based studies reveals that, in 21 of these, researchers are merely observers of birds, much like their nineteenth- and early twentieth-century counterparts, the only difference being that p-values have replaced percentages. For instance, a study of nest competition between starlings and woodpeckers merely records which species ultimately occupies saguaro nest cavities:

> We intensively searched each plot for several days until we were sure we had located all the European Starling, Gila Woodpecker, and Northern Flicker nests. The visibility in saguaro cacti forests allowed us to see birds fly to and from their nests, and we could often hear nestlings calling from a nest. . . . When we were uncertain if a cavity was a nest, we climbed the saguaro with a ladder and looked into the cavity with a mirror and light. If we could not reach a suspect nest cavity with a ladder, we checked the cavity every week until nestlings could be heard, a bird was observed feeding nestlings, or the nesting season ended. (Kerpez & Smith, 1990, p. 368)

But, as Figure 5.10 illustrates, the article is not merely a narrative description of their observations. It is instead permeated with p-values.

Another article describes dominance among Lesser Snow Geese by observing the birds' interactions. One study of Peregrine Falcon predation simply counts the number of corpses found, whereas another researcher counts the number of seeds consumed by crossbills. Still another records whether Snail Kites eat turtles, crabs, or snails.

In fact, only nine of the 65 articles in the 1990 volume fail to reference statistics. Of these, two are theoretical, three use advanced laboratory technology, one uses elaborate mathematical formulas, another is a Japanese behavioral study, and a final one is a museum specimen study.

The remaining two articles, however, are noticeably different. One of these, "Immediate Impact of the 'Exxon Valdez' Oil Spill on Marine Birds," (Piatt et al., 1990) is essentially a form of natural history discourse, giving a descriptive account of the Exxon disaster. The results section is merely a narrative account of the incident, though the authors use elaborate means to gather data (ship and aerial surveys) and to reference laboratory procedures (autopsies and computer-aided estimates) in order to estimate numbers of birds killed. Its acceptance here is likely due to its currency, the ecological importance of the event narrated, and the study area's inaccessibility.

The second article shows that, even in 1990, amateurs can still play a part in ornithological science. The methodological section of "Migration Routes and honey mesquite (*Prosopis juliflora*), ($r=0.5333$, $P = 0.041$). Together, the year the plot was censused and the volume of ironwood explained 45.7% of the variation in the number of Northern Flicker nests ($P = 0.0257$, intercept = 1.958).

There were no significant differences in the number of Northern Flicker nests on plots with different aspects ($P = 0.726$).

Factors affecting the number of European Starling nests.—The mean number of European Starling nests on the plot with European Starlings was 3.00 ± 0.926 ($n = 7$). The distance from the plot to the nearest agriculture was negatively related to the number of European Starling nests and alone explained 29.2% of the variation in the number of European Starling nests ($P = 0.0078$, $b = -0.2260$). The mean date the plot was censused was negatively related to the number of European Starling nests and explained an additional 26.3% of the variation in the number of European Starling nests ($P = 0.0208$, $b = -.02319$). Together, these two variables explained 55.5% of the variation in the number of European Starling nests ($P = 0.0078$, intercept = 2.220).

Plots with northern aspects had significantly fewer European Starling nests than did plots with eastern aspects ($P = 0.027$). All three plots with northern aspects had no European Starling nests; however, this was probably not caused by their northern aspect but instead by their distance from agriculture. The three plots with northern aspects were all > 4 km from agriculture. The farthest plot from agriculture with a European Starling nest was ca. 4 km from agriculture.

There was no significant difference between 1984 and 1985 in the number of European Starling nests on the four plots censused both years ($P = 0.392$).

Nest-cavity dimensions and location.--Throughout the study, 64 Gila Woodpecker, 28 Northern Flicker, and 26 European Starling nests were found on or near the plots. We measured the cavity dimensions of 32 Gila Woodpecker, 15 Northern Flicker, and 19 European Starling nests. European Starlings and Gila Woodpeckers did not significantly differ in nest-cavity dimensions ($P = 0.3840$). European Starlings and Gila Woodpeckers both significantly differed from Northern Flickers in nest-cavity dimensions ($P = 0.0182$ and $P = 0.0001$, respectively.) European Starling and Gila Woodpecker nest cavities had significantly smaller entrances (vertical and horizontal diameters), and the cavities were significantly shallower in the vertical plane of the saguaro than Northern Flicker nest cavities (Table 2). The three species did not significantly differ in nest-cavity height (Table 2).

European Starlings and Gila Woodpeckers did not differ significantly in their use of saguaros for nest sites ($P = 0.511$) (Fig. 2). European Starlings and Gila Woodpeckers nested in class 3 saguaros significantly less and in class 4 saguaros significantly more than Northern Flickers ($P \leq 0.013$) (Fig. 2). The orientation of nest cavities did not significantly differ from random for Gila Woodpeckers ($r = 0.02$, $P > 0.90$), Northern Flickers ($r = 0.16$, $P > 0.47$), and European Starlings ($r = 0.21$, $P > 0.30$). For more information on Gila Woodpecker and Northern Flicker nest-site selection and nest-cavity characteristics see Kerpez and Smith (1990).

DISCUSSION

Competition between European Starlings and Gila Woodpeckers.—European Starlings compete with Gila Woodpeckers for nest cavities in saguaros and this competition decreases the number of

FIGURE 5.10. Text page demonstrating an observation-based study permeated with statistical references (Kerpez & Smith, 1990, p. 371). Used with the permission of the American Ornithologists' Union.

of New World Sanderlings (*Calidris alba*)" describes in detail the use of birders' records to conduct its study:

> From the literature and from knowledgeable birders, we identified a series of locations around the U. S. likely to be major passage points for Sanderlings in migration. . . . We built a network of volunteers to survey likely Sanderling migration stopovers by soliciting for collaborators in popular articles, semiregular newsletters, and widely distributed posters. . . . When the volunteer network identified local concentration areas, we visited and personally surveyed Sanderlings for marked birds (Myers et al., 1990, pp. 173–174).

In all, volunteer collaborators reported 55% of the 252 independent reports; of these, 41% were from "non-professional volunteers" (p. 174), the rest from "people actively involved in shorebird research" (p. 174). Although American ornithology in 1990 still permits this strictly natural history discourse and direct participation by amateurs, *The Auk*'s editor places both of these articles at the end of their respective sections—that is, 17th out of 17 articles and 19th out of 19 respectively, perhaps indicative of their place on the periphery of ornithological science and thus the clear predominance of experimental science discourse.

Organization and formatting

During this most recent phase in the development of American ornithological discourse, the journal itself underwent two changes in format. In 1977, pages had become 3/4" x 1" larger; and beginning in 1982 a double-column page layout was used, thus increasing text density. Also in 1982, "Information for Contributors" had expanded to three full pages, with a page and a half on manuscript preparation. Instead of sentence-length descriptions, there were now entire paragraphs on manuscript format and style, the writing of abstracts, scientific nomenclature, use of mathematical and symbolic units, citations, tables, illustrations, drawings, and photographs. Also during this period, a standard format for articles evolved, reflecting the establishment of experimental science discourse in American ornithology. Added evidence of the changing nature of American ornithology is the small number of contributing authors not stating professional affiliation, only 10 of 175, or 6% unaffiliated.

In the 1970 volume, all but one of the authors used at least some of the standard headings to subdivide their articles. The Summary, Acknowledgments, and Literature Cited sections were almost uniformly used, though other section headings did appear frequently also. In addition, ornithologists continued to use other descriptive headings to introduce sections.

By 1980, the Summary at the end had been replaced by an Abstract at the beginning of each article. As Davida Charney (1993) discovered during her protocol analysis of reading patterns of evolutionists, these initial abstracts have become essential for the busy researcher (pp. 211–212). In addition, all authors were using some form of standard headings, and Acknowledgments were uniformly used. Even two articles patterned after Figure 5.2 above, the century-old bird list, had now added Acknowledgments and Literature Cited sections. Articles in the first half of the 1980 volume often still added original headings; and others wrote variant forms of the standard ones, like *Results and Discussion*, *Discussion and Conclusions*, or *Materials and Methods*. By the publication of the last issue in 1980, however, most original headings had disappeared, and the standard format of Abstract, Methods, Results, Discussion, Acknowledgments, and Literature Cited was now in place. And in 1990, though there was still the occasional variation, this format was firmly established as the norm for American ornithological discourse.

Graphics

This final period, 1970–1990, brought with it a tremendous increase in the amount and diversity of graphics in American ornithological discourse.

In all, there was an almost four-fold increase in the use of graphics. The number of formal tables tripled, and informal tables had completely disappeared. More importantly for experimental science discourse, there were eight times as many graphs, with line and bar graphs and scatterplots the most prominently used. As ornithology specialized, so did its graphics which employed new forms such as phenograms, sound spectrograms, and cladograms. Graphics were no longer simple presentations of one set of data but were now used to summarize large quantities of the empirical data ornithologists were gathering in the field and lab. There was not just one bar graph to a figure, but five or 10. Hybrid graphical displays occasionally appeared using more than one type of graph per figure; an extreme case is shown in Figure 5.11, which includes bar graphs, histogram, line graph, and scatterplot stacked and interwoven into one figure.

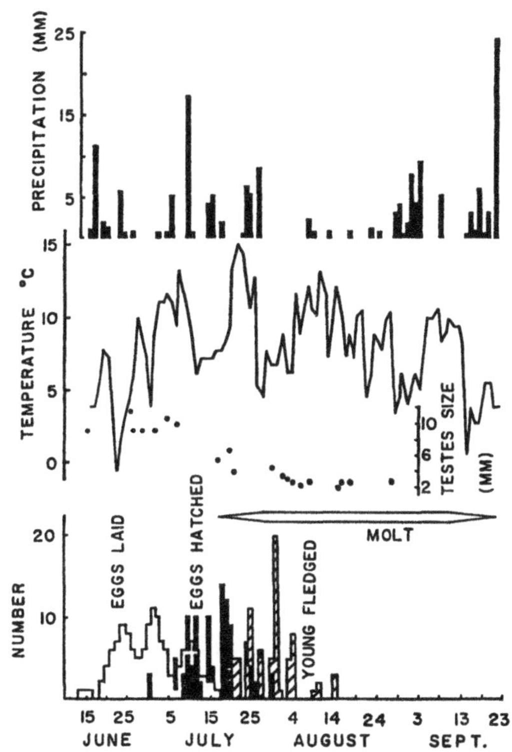

FIGURE 5.11. An extreme case of graphical display during American ornithology's evolution to experimental science discourse (Verbeek, 1970, p. 445). Used with the permission of the American Ornithologists' Union.

In fact, the editors of *The Auk* encouraged the use of illustrations such as these. Beginning with the 1977 issue, *The Auk*'s "Information for Contributors" commented that illustrations are "one of the most effective ways of summarizing or emphasizing the findings or ideas of a study" (Information, 1977, p. 192). This suggestion perhaps also contributed to the continued use of drawings. However, photographs were declining in popularity, following the route of natural history discourse. The big losers were lists, which had almost disappeared along with the bird lists and specimen studies popular during the three previous periods of discourse. During this fourth period, the presence of complex graphics—such as cladograms, sound spectrograms, DNA fingerprints, principle component diagrams, and scatterplots with their accompanying regression lines—signals the predominance of experimental science discourse in American ornithology. By the frequency of their use, these complex graphics have themselves been accomplices in the break with natural history discourse by excluding the amateur through sheer complexity of meaning. However, the editors of *The Auk* encouraged their contributors to keep graphics as straightforward as possible, as shown in the following comment first made in the 1982 "Information for Contributors": "[Tables] should generally be small, simple, and easy to read, and should communicate a clear message" (Information, 1982, p. 185).

Co-authorship, Article Length, and Volume of Publication: 1884–1990

The rise of experimental science discourse also signaled an increase in the amount of collaborative articles published by American ornithologists. Although an occasional two-authored article was common throughout *The Auk*'s history, co-authored articles began to appear with more frequency in the 1940, 1950, and 1960 volumes, the period of discourse interplay, when experimental science discourse was beginning to take hold. The 1970 volume, which this study has shown to signal the transition to experimental science discourse in *The Auk*, registered a noticeable increase in co-authored articles. A third of all articles published in 1970 were co-authored; and, in 1980 and 1990, the percentage of multi-authored articles had risen to over 60%. In these last two volumes, co-authored articles were the most common; but articles written by three, four, and five ornithologists constituted at least a fourth of all articles published by *The Auk*. Extreme cases of co-authorship were two articles carrying 10-author bylines.

Article length and number of articles per volume also show reversals of trends during the third phase of American ornithological discourse, as shown in Figure 5.12. The number of articles published yearly by *The Auk* declined until 1960, whereas the number of pages per article generally increased until that same year.

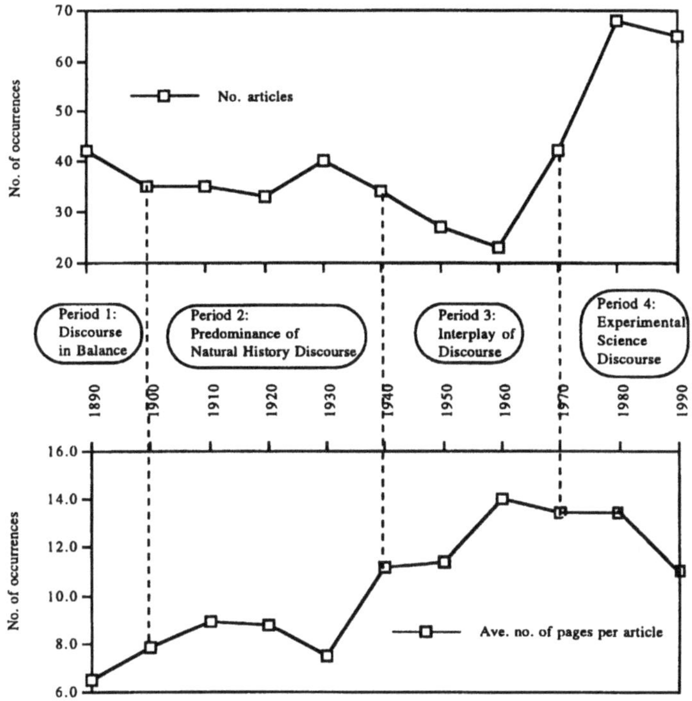

FIGURE 5.12. Changes in number and length of articles (N=444).

However, the top graph of Figure 5.12 needs some interpretation because the amount of space devoted to articles fluctuated over the life of the journal. The percentage of change between the number of articles published and the space given to all articles in *The Auk* starting with the 1890 volume and each 10-year interval thereafter is as follows:

Year	*% Change over Previous Vol. No. of Articles*	*% Change over Previous Vol. Total No. Pages for Articles*
1890	—	—
1900	(17)	2
1910	0	12
1920	6	(7)
1930	21	3
1940	(15)	27
1950	(21)	(19)
1960	(15)	5
1970	**83**	**74**
1980	**62**	**62**
1990	**(4)**	**(21)**

It is apparent that the drastic increase in the number of articles published beginning with the 1970 volume was more a function of the increased space allocated to articles in *The Auk*, rather than to the decreasing length of individual articles shown in the bottom graph of Figure 5.12. To help compensate for the increase in the number of articles, less space was allocated to the rest of the journal (14% less in 1970 and 34% less in 1980), a fact that emphasizes even more the increase in article publication. That is, by 1970 the journal began devoting much more space to full-length articles, perhaps indicating an awareness of the increasing professional nature of the organization.

Regarding the reversal in the length of published articles during the period of discourse interplay (Figure 5.12, bottom graph), editorial decisions by the journal itself do not seem to be the impetus for decreasing article lengths. Information for contributors printed in the 1950 through 1961 volumes, when article lengths began to decline, mentioned only that "long articles should be summarized briefly" (*Auk*, 1950). In January 1963, the term "long articles" was defined as "more than 10 pp" (Suggestions). Finally, in order to distinguish articles from Short Communications, a January 1977 "Information for Contributors" defined articles as contributions that "generally exceed 2,500 words" (Information, 1977, p. 191)—that is, about four full pages of text in the 1977 volume, far below the average number of pages throughout *The Auk*'s history. But in none of these editorial descriptions between 1950 and 1977 was there an attempt to restrict page limits. The need for authors to consider the page length of manuscripts does not appear until 1978, long past the 1960 reversal. Perhaps decreases in article lengths during this time reflect the increase in graphical display (see below), which displaces the need for elaborate description.

After 1977, however, editorial decisions do hint at the increasing pressure on ornithologists to write more concisely. A policy statement in the February 1978 issue noted the increasing discrepancy between members' dues and publication costs, adding that "authors of longer papers are especially encouraged to help defray the costs of publication; it is unlikely that papers of more than 16 printed pages can be published without some significant contribution" (Policies, 1978, p. 219). Another indication of the journal's attention to article length appeared in an *Auk* editorial 7 years later in which editor Alan H. Brush (1985) commented that acceptance of articles was in part a function of "page limits" (p. 903). As Myers (1990) has noted, most academic journals allocate space to an article by the perceived importance of the author's claim, rather than by prescriptions in "Instructions to Authors" (p. 86).

It should be noted here that, in 1977, pages became slightly larger (3/4" x 1" larger), and in 1982 the journal changed to a double-column layout. These changes raised the average number of words on a full page of text from about 625 to 750 words, or a 20% increase in words per page. Thus, in the tables, graphs, and text discussion of this chapter, calculations of page lengths for articles published after 1976 have been adjusted by 20% to account for these formatting changes.

Changes in Graphical Presentation: 1884–1990

As the chronology of historical change in this chapter has shown, the use and design of graphical displays have likewise changed drastically over *The Auk*'s history. Figure 5.13 highlights the most significant changes revealed by the data[1]. Though graphics were used from the beginning of the journal's history, their use increased exponentially during the fourth period of American ornithological discourse history, doubling in 1970 and nearly quadrupling by 1990. The use of photographs, however, has had an

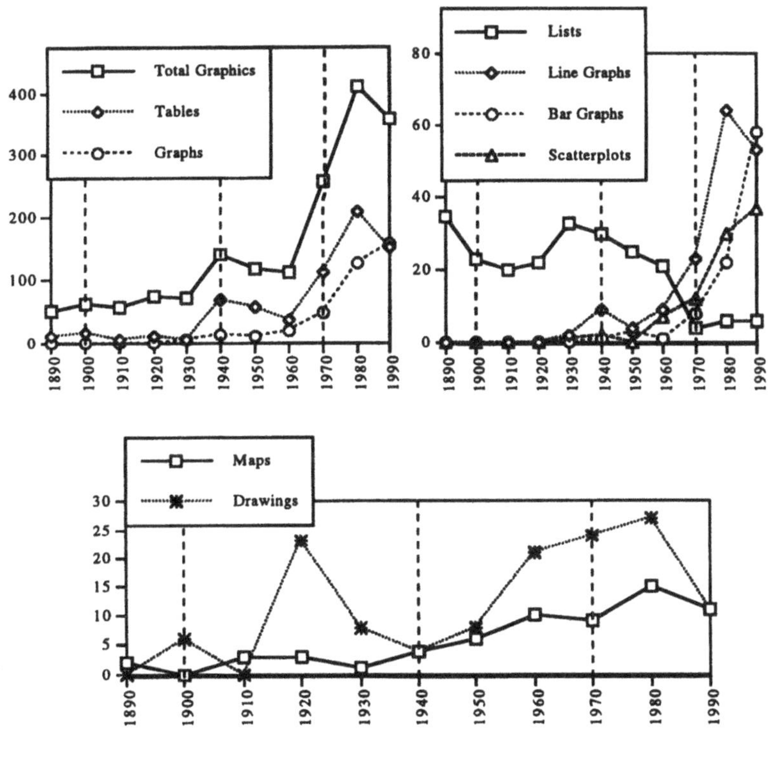

FIGURE 5.13. Chronological changes in ornithologists' use of graphics (Y-axes record the total number of occurrences per 10-year volume of *The Auk*. N=444).

erratic history, declining along with natural history discourse but reemerging in 1970. Tables and graphs are visual markers of experimental science discourse, as their use increased greatly toward the end of the period of discourse interplay when experimental science discourse began to dominate. Both were the preferred graphic, used almost equally by ornithologists by the 1990 volume. Formal tables superseded informal ones, which disappeared during the period of discourse interplay between 1940 and 1970.

As shown in Figure 5.13, line and bar graphs were most preferred, though the use of scatterplots rose more rapidly during the transition to experimental science discourse during the 1960s and 1970s. Lists, both numbered and unnumbered, were the clear losers in the transition to scientific discourse, almost disappearing by 1970. The use of drawings and maps was less affected by the changing discourse, both registering more modest, though generally continued, growth.

LINGUISTIC ANALYSIS OF ORNITHOLOGICAL DISCOURSE

So far, this historical study of *The Auk*, American ornithology's major journal, has demonstrated the rhetorical contexts in which the American ornithological community has operated. Since the journal is the principal American publication in the field of ornithology, a study of its discourse should also reveal the linguistic decisions made by American ornithologists during the past century. These decisions should mirror social interactions both within the community and with outside forces.

The next component of this study consists of a linguistic analysis of a systematic sampling (see Cochran, 1977) of 44 articles from the journal since its inception in 1884[2]. Once occurrences were counted, the data was grouped into 10-year periods, averaged, and graphed as frequency distributions for evidence of change throughout the journal's history.

Method for Analysis

The analysis of 1,000-word passages from the database focused on the same features of discourse used in chapters 3 and 4 to describe the three discourse models used in this study of *The Auk*:

- the use of multi-word noun phrases, technical terminology, nominalizations, and definition
- the use of metadiscourse
- sentence analysis based on the following features: average sentence length; the use of personal pronouns and passive voice; verb usage; and

sentence complexity, shown by the occurrence of subjects as placeholders, use of complex clauses, and interruptions of subject and verb
- the use of evidentials

Transforming the data into time-line graphs will show positive or negative trends in these 32 linguistic features. These patterns may then be correlated with the historical changes from natural history to experimental science discourse already described in chapter 4. In interpreting these linguistic features, it is necessary to look at only obvious, decided changes in the slopes of the time lines since it is impossible to separate completely an individual writer's impact upon any piece of discourse, as has already been discussed in chapter 3. An individual, one-period irregularity (that is, outlier) may be caused by one particular author's overuse of the linguistic feature being sampled. See for instance the data for personal pronouns shown in Figure 5.14. Despite this drawback, those time lines that show decided slopes may be even more significant, because they are showing changes despite the limitations inherent in such a study. These lines most likely point to those linguistic features indicative of the most change over time and are consequently the most useful for distinguishing natural history and experimental science discourse. Further study of more articles might show even larger changes and steeper slopes.

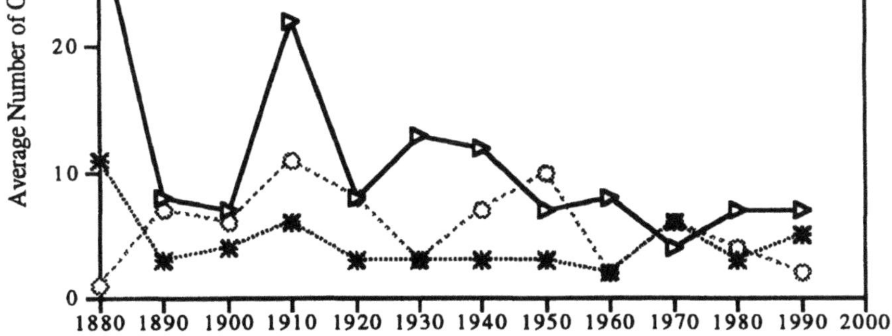

FIGURE 5.14. Example of irregularities in data likely associated with individual linguistic variation (n=44).

Two Perspectives in Analyzing Data

Before proceeding in the analysis, some of the linguistic features must be discarded because there is not enough sample data for analysis. That is, for some features there were on average no more than a few examples for each 10-year period. Eliminated then from further analysis because of a lack of data are the three types of metadiscourse, use of definitions, and four types of evidentials (belief, deduction, hearsay, and induction). The 24 remaining linguistic features may be studied from two viewpoints:

- linguistic features that are generally stable throughout, and
- those that show definite changes over time.

Stable Linguistic Features in a Changing Ornithological Discourse

Seven of the linguistic features are generally stable throughout the century of sampled discourse; that is, values early in the century are similar to those late, and there are no large positive or negative slopes, though there may be some fluctuation likely due to individual variation. Three of these features (use of personal pronouns and reporting verbs and reliance on hedging to make factual claims) have been used consistently over time in American ornithological discourse. The persistence of reporting verbs may be explained by the pervasiveness of observation-based discourse throughout American ornithology, which requires such verbs. Though the presence of hedges (such as *about* or *approximately*) may appear incongruent with later experimental science discourse, ornithology is perhaps unique among scientific disciplines in that it regularly reports, for instance, migration and population estimates of species, which, with the exception of data collected by sophisticated radar equipment, must be reported only in approximations. Perhaps unusual is the persistence of personal pronouns despite the rise of experimental science, the discourse of which is traditionally noted for its impersonality and appearance of objectivity. Personal-pronoun usage may be one of the linguistic features that set American ornithological discourse apart from other scientific disciplines because of the continued presence in American ornithology of amateur discourse, the authors of which do not have agendas as concerned about perceived objectivity.

Though sentence complexity and length do vary over time, as will be shown later, four clausal features (relative noun clauses and subordinates of time, cause/effect, and condition) remained relatively stable. Because a popular topic for research throughout ornithological history has been population dynamics (that is, changes in bird populations over time), stability in the use of subordinate time clauses is understandable. However, the consistency in usage of the other three clauses may be unexpected. Surprising also is the general stability of causal and conditional clauses, despite the increase in predictive and theoretical issues in recent ornithology, as shown earlier in this chapter.

Changing Linguistic Features in a Changing Ornithological Discourse

The focus here is on those linguistic features that show definite changes over time (that is, those features with definite upward- or downward-sloping time lines). These features are the most valuable data because changes in their usage over time suggest a relationship with the historical changes from natural history to experimental science discourse as described earlier. It will be proposed in chapter 6 that these selected features may be considered *discourse indices* of the professionalization of a science; that is, their presence in a particular piece of writing may identify that discourse as either natural history or experimental science discourse.

For purposes of analysis, the 16 linguistic features showing variation are grouped here into three categories: eight dealing with sentence complexity and four each influencing vocabulary usage and making factual claims. These are then subdivided further in order to explain variations by showing how features interact among one another. To aid in the analysis of these changes, vertical lines are added to the graphs below, dividing time lines into the four historical periods discussed in the first half of this chapter:

1884–1900: The Discourses of Natural History and Experimental Science in Balance
1900–1940: The Predominance of Natural History Discourse
1940–1970: An Interplay of Discourse
1970–1990: The Emergence of Experimental Science Discourse

Sentence complexity

Variations in sentence complexity are demonstrated in the following three graphics showing changes in sentence length, subject-verb interruptions, clausal variations, and use of impersonal subjects. Figure 5.15 shows a marked increase in the number of simple sentences and a corresponding, though less pronounced, decrease in average sentence length. Though it may be expected that scientific discourse increases in complexity as science increases in sophistication and complex argumentation, the early literary quality of American ornithological writing due to the influence of amateurs in natural history discourse tended to produced more lengthy, intricate sentences. For American ornithology, at least, increased sentence simplicity occurred in conjunction with increased professionalization; that is, the discourse became "less literary." In addition, historical factors, such as shifts in literacy and literary taste, played a role in the shifting ornithological style.

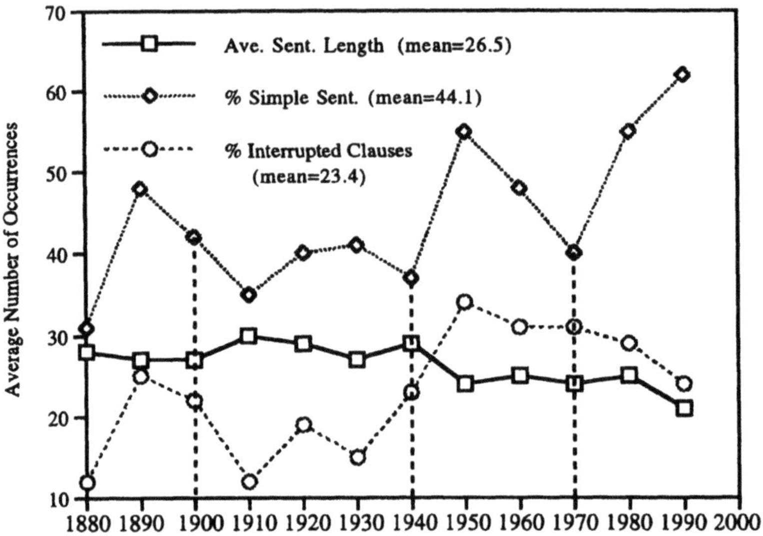

FIGURE 5.15. Average changes in sentence length and complexity (n=44).

Figure 5.15 also shows the most striking historical change in the linguistic features studied. The number of clauses containing interruptions of subject and verb increases from between 10 and 20 percent over the course of American ornithological discourse history. Though the sample data also indicate that the increase is accompanied by a three-fold increase in the number of interrupting words separating subject and verb as ornithological discourse becomes more professional, sample size and an unusually large variation in this data may be exaggerating the effect. Regardless, a decided increase in the number of clausal interrupters themselves is clearly another benchmark in the switch to experimental science discourse, as evidenced particularly by the sharp increase during the third period of discourse interplay.

Decreases in usage of relative adjective and subordinate place clauses shown in Figure 5.16 parallel increases in sentence simplicity described earlier. The decrease shown in Figure 5.17 in the use of subjects as placeholders (such as *it* and *one*) may be influenced by the increase in the use of passive voice, which places objects in subject position.

Vocabulary

Since the professionalization of science is accompanied by an increase in expertise and greater theoretical sophistication, changes in the vocabulary of scientific discourse should be predictive indices of an increasing professionalism. Such are the next four linguistic features of

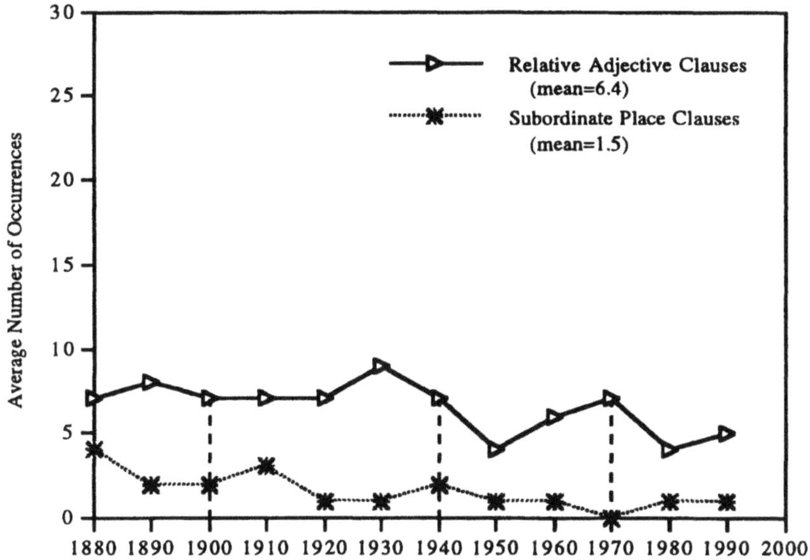

FIGURE 5.16. Average changes in two clausal features (n=44).

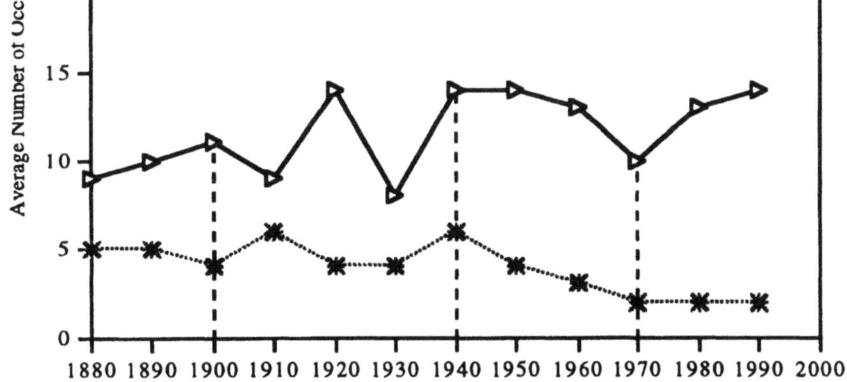

FIGURE 5.17. Average changes in the use of impersonal subjects (n=44).

American ornithological discourse. Figure 5.18 shows a gradual but decided decrease in the use of *be* verbs by American ornithologists. Accompanied by a corresponding rise in substantive verbs, the decrease in *to be* usage, despite the rise of passive voice, indicates increased sophistication of expression by ornithologists. There was, however, an attempt by the journal beginning with the January 1977 "Information for Contributors" to minimize complexity of expression, at least in the Abstract: "[In abstracts] use active verbs, and emphasize ideas" (Information, 1977, p. 191).

Given previous studies of the language of scientific discourse, the marked increases in American ornithologists' use of nominalizations and technical terms as shown in Figure 5.18 are no surprise. These expected results, however, do add credibility to the rest of the data gathered in this study. Significant here is the rapid increase of all three linguistic features during the third period of American ornithological discourse change, when the discourse was being redefined for use by a rapidly developing scientific profession. Consequently, the presence of nominalizations and technical and polynomial terms can be considered three more benchmarks, along with subject-verb interruptions, as suggested in the previous section, for distinguishing discourse models.

Evidentiality

Besides changes in vocabulary usage, the rise of professionalism in science is also accompanied by changes in argumentation. A study of the use of

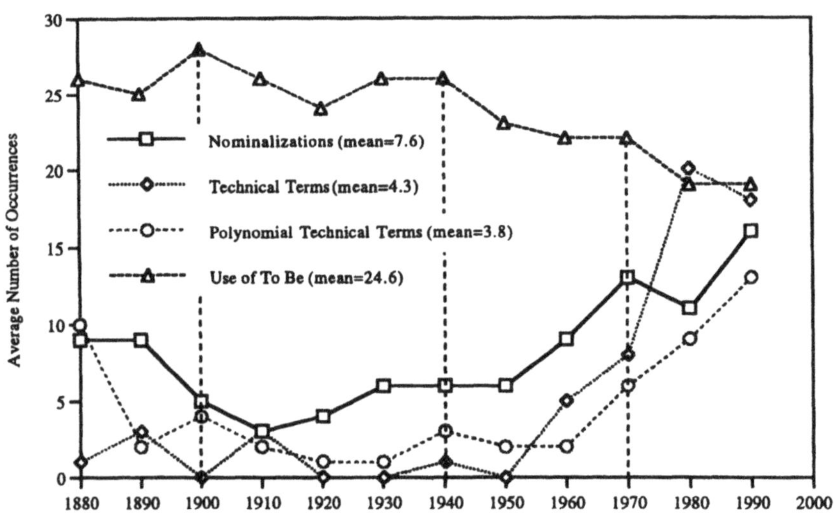

FIGURE 5.18. Average changes in vocabulary (n=44).

graphics, described earlier in this chapter, is a visual way to illustrate such changes. A linguistic method for documenting these same changes is the variation in the use of evidentials, or markers for making factual claims. As shown previously, the increased sophistication of computer technology has given rise to statistical inference in American ornithology. It would be assumed that the universally accepted method of scientific argumentation by means of hypothesis testing and inferential analysis would replace the need for linguistic aids for making claims. Figure 5.19 appears to verify this assumption, as illustrated by the decrease in usage of sensory evidence (that is, usage of *see*, *hear*, *feel*, etc.) and markers of expectation (that is, *of course*, *in fact*, *even*, etc.).

Not surprising either is the sharp increase since 1920 in the use of citations by American ornithologists, another of the benchmark features identifying change in scientific discourse (see Figure 5.20). Less clear is the gradual fall, then more rapid rise, in the use of markers indicating the reliability of evidence (like *maybe*, *perhaps*, *generally*, *virtually*, etc.). It would seem that such markers should continue to decrease as scientific argumentation becomes more sophisticated. Here, the decline is reversed during the third period described as discourse interplay. Perhaps the reversal in usage of markers of reliability of evidence is not caused by discourse change but rather by the fact that statistical inference is based on probability, which in turn has reintroduced this marker into American ornithologists' vocabularies.

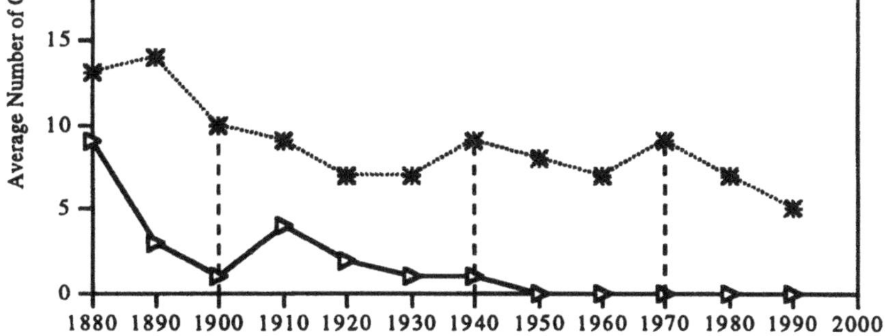

FIGURE 5.19. Average decreases in evidentiality (n=44).

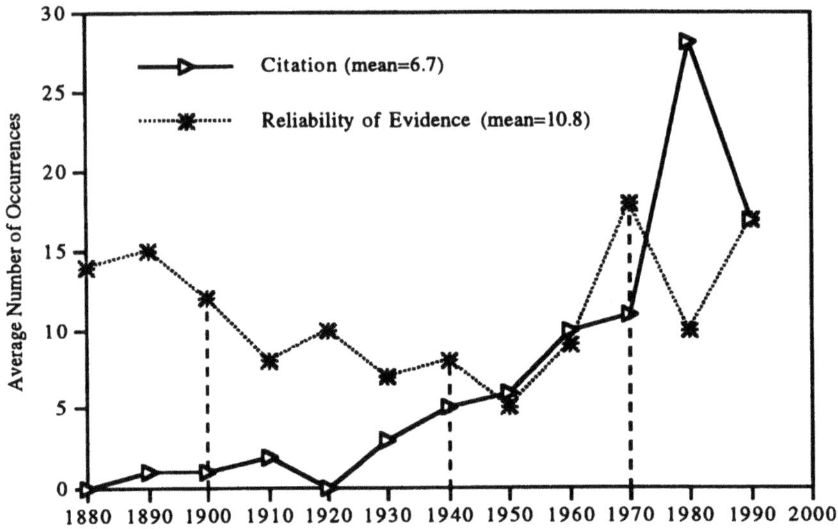

FIGURE 5.20. Average increases in evidentiality (n=44).

QUESTIONS FOR FURTHER STUDY

Chapter 4's overview of 444 articles, which span the existence of the American Ornithologists' Union, showed the 100-year transition to experimental science discourse. This chapter has delved more deeply into this century of ornithological discourse, proposing four periods into which the discourse falls: discourse in balance before the turn of the century, preponderance of natural history discourse during the first 40 years of the century, an interplay of natural history and experimental science discourse during mid century, and the emergence of experimental science discourse after 1970. Reflecting these changing periods, organization and formatting of ornithological discourse also became standardized in the 1980s with the establishment of experimental science discourse. By the end of the twentieth century, *The Auk* was publishing more, though shorter, articles. Data display became more varied and complex through the changing periods, demonstrating the continuing sophistication of American ornithological discourse and its more exclusively professional audience. The second half of the chapter sought linguistic indices to reflect the changing discourse models. Serving as the best benchmarks indicating the presence of experimental science discourse are significant increases in the percentage of interrupted clauses, of nominalizations and technical and polynomial terms, of markers for reliability of evidence, and of the use of citations. Despite the increasing theoretical complexity and push for scientific objectivity, the use of personal pronouns, subordinate causal and conditional clauses, and hedging devices remained relatively stable, perhaps reflecting the prolonged interrelationship between the amateur and professional.

What are the reasons for this persistence of amateur discourse in ornithology? Is the apparently simple, straightforward method of observation in many ornithological research studies enough like natural history observation to permit natural history discourse to survive? That is, despite limitations due to experimental design, restricted access to research sites, and sophisticated instrumentation, is the way the discourse reads necessarily the same because writers can linguistically describe what they *see* in only a limited number of ways?

Many other questions arise from this 100-year survey of *The Auk*. Is the apparent shift to theoretical science after 1970 caused merely by the appearance of statistical inference or has something happened to natural history study as well? Why was there the surge of interest in natural history around 1910? Why does natural history discourse decline in 1930 and revive in the next three volumes? If ornithological science is professionalizing, why is natural history discourse able to maintain its ground between 1940 and 1970? For the science of American ornithology, has professionalization not meant specialization and exclusion? How does the discourse change in order to present such controversial subjects as DNA recombination? Though the number of articles increases over the 100-year period, why does the length of articles decrease, unlike the discourse of other scientific fields like astronomy and physics? Does the progression of American ornithological science actually fit into the Continuum proposed by Jimmie Killingsworth and Jacqueline Palmer (1992), or should the Continuum be modified for American ornithology? Chapter 6 will consider these questions.

6

Scientific Discourse From the Perspective of American Ornithology*

This final chapter brings together the historical, social, linguistic, and discourse components of American ornithology discussed in the past four chapters into a unified picture of the forces affecting American ornithologists' writing over the past century. The discourse models described previously in chapters 3 and 4 are also refined further. The study ends by fulfilling its ultimate goal of placing American ornithological discourse in the larger context of scientific writing as it has been defined to date.

Of particular importance for this study, Charles Alan Taylor (1996) has asserted the value of historical narratives in rhetorical analysis of science, including its professionalization. Like Charles Bazerman (1988) and Greg Myers (1990), he has argued well for "close textual studies of science and scientific practice" (p. 81). This present study has focused primarily on written texts, which, as Myers (1990) pointed out, are valuable objects for study because they are concrete instances of communication that can be analyzed and re-analyzed

* Part of this chapter has appeared previously in Battalio, J. T. (1996). The interplay between narrative, education, and exposition in an emerging science. *Journal of Technical Writing and Communication, 26,* 177–191. © 1996, Baywood Publishing Co., Inc.

in depth (pp. 5–7). They can be reinterpreted and can guide other interpretations, as additional data and theories arise. Perhaps the most valuable contribution that a study of American ornithological discourse can make to the question "What is scientific discourse?" is a clarification of what it means for science writing to advance from a primarily amateur to a predominantly professional discourse. The scientific discourse studied by previous analysts, as for instance physics and astronomy, was written by scientists who had a clear understanding of what it meant to be an astronomer or physicist. However, the question of what it means to be an "ornithologist" has been answered only relatively recently, because the value of the *scientific* study of birds has become clear only during this century. Consequently, a study of American ornithological discourse can, in the space of only 100 years, trace the science from its inception, through the realization of what it meant to be a professional ornithologist.

An additional benefit from this study of American ornithological discourse is the ability to define two poles of discourse in science, which this study has defined as natural history and experimental science discourse. Viewing ornithological discourse as a professionalizing discourse adds a third perspective, a transitional discourse model. In defining the boundary between amateur and professional ornithological discourse, we must, above all else, determine the essential characteristics of science writing. That is, what features distinguish professional science writing from its amateur counterpart? The rapid changes in ornithological discourse during the mid-twentieth century in the use of linguistic features such as technical terminology, nominalizations, and subject-verb interruptions; in the presentation of complex visual displays; in the volume, quantity, and co-authorship of publication; in the means of presenting evidence; and in the percentage of content devoted to preparation, process, and production of research show that writing does not become scientific until certain conditions are reached. At what point do these changing features merge to create a professional discourse? Is there no hybrid science writing? Interrelating the characteristics of American ornithological discourse and its history with those of the discourses of physicists, astronomers, and ecologists already described by researchers and rhetoricians will give further insight into the answers to these questions, thus fulfilling the ultimate aim of this study: to improve communication by improving our understanding of the social and historical forces that affect scientific communication. In doing so, this study employs both historical and contextual analyses of scientific discourse and practice. M. A. K. Halliday and J. R. Martin (1993) have also argued in favor of a multi-level analysis of science writing: "It is only by shunting between language and social context (that is, between the planes of science as text and science as institution) that we can begin to map out a meaningful interpretation of the discourse of science" (p. 26).

A CLOSER LOOK AT THE FOUR PERIODS OF ORNITHOLOGICAL DISCOURSE

This study has proposed that the history of American ornithological discourse be divided into four significant periods.[1] The first, 1884 to 1900, was a relatively quiet time for the developing discourse of ornithologists when there appears to have been a balance between writing characterized as either natural history or technical ornithology. Then, due to an apparent redefinition of American ornithology during the early part of the twentieth century, natural history discourse became the dominant feature of *The Auk*. This second period of discourse, from 1900 to 1940, recorded the rapid rise and, by the end of the period, decline of strictly natural history writing by ornithologists. During the post–World War II period from 1940 to 1970, there were significant changes in most aspects of American ornithological discourse, everything from linguistic features, such as sentence lengths, diction choices, and grammatical usage; volume of publication; amount of coauthorship; formatting; use of visuals; descriptions of methodology; and means of argumentation. As a result, during the last two decades, a professional discourse of experimental science has emerged, one that signals the professional maturation of American ornithology. It is now time to take a closer look at these four periods in light of other historical and social events taking place at the same time.

Conditions in Late Nineteenth-Century Ornithology

It has been argued that, by the late nineteenth and early twentieth centuries, most scientific fields had defined clear boundaries between amateur and professional (Daniels, 1967; Holmfeld, 1970; Reingold, 1976; Larson, 1977; Turner, 1978; Shapin, 1982, p. 173). This was certainly not the case for American ornithology. Because of the union of amateur and technical aficionado of birds into a national organization in 1883, both maintained a close association during this first period, sharing the pages of *The Auk* to describe their ornithological study. Thus, both amateur and technical discourse appeared side by side. Through this new journal, members were in effect describing what it meant to be an American ornithologist.

There were also explicit signs that the young scientific community was beginning to organize itself into a profession. By the late nineteenth century, several reference texts, manuals, and field guides about birds had appeared, including Elliott Coues' *Key* (1872), American ornithology's first textbook. Writers of the period began using the term "technical ornithology" to describe specimen collecting and museum cataloging. The term is the most evident attempt of early ornithologists to distinguish themselves at least linguistically from the casual observer of birds or practitioner of natural history.

Another early attempt to distinguish amateur from technical ornithologist occurred in the late nineteenth century as American ornithologists split into two camps over the inclusion in the ornithological literature of the technical descriptions of the birds discovered by eighteenth-century travel writer William Bartram. Daniels (1967, p. 155), for instance, considered conflicts over nomenclature to be an early indication of professionalization. J. A. Allen, then president of the AOU, led the assault to disavow much of Bartram's work by removing from the ornithological literature Bartram's type descriptions and nomenclature. On the opposite end was systematic ornithologist Elliott Coues, who staunchly defended Bartram's binomial names for the birds the botanist discovered. Coues' defense was to no avail, however; for he was forced to offer alternative scientific names to replace the last of Bartram's nomenclature, but not before registering a final protest in *The Auk* (see Coues, 1899). About all that remains today in the ornithological literature of Bartram's influence is the scientific name for the Upland Sandpiper (*Bartramia longicauda*), one of the species first discovered by Bartram and described in both his journal notes and his *Travels*. Though ornithologists point to inconsistencies in Bartram's binomial nomenclature for this rejection (Stevens, 1936, p. 204), at work also is ornithologists' desire to distance themselves from Bartram, the naturalist. His was a personal narrative freely interspersed with nonscientific anthropomorphisms and religious emotionalism for the glory of God and nature, themes particularly representative of the amateur and antithetical to the drive for professionalization.

But "technical" ornithologists of the period had a difficult time separating themselves from the amateur. The personal narrative was a key genre in early American ornithologists' writing, with writers describing the location, sighting, and physical description of the birds they observed. Even in those technical studies that presented detailed studies of specimens in elaborate table formats, the data used for many of these early *Auk* articles were very easily obtainable simply by receiving permission to visit a museum where the desired specimens were stored. Likewise, the almost complete absence of citations in the early *Auk*, except for ornithologists referencing their own field notes, reflects not only the absence of a body of work containing accepted paradigms, but also the lack of a recognized core group of experts to oversee the developing scientific community.

And because scientific rhetoric requires that its legitimate assertions be authorized, they must be addressed to specialists who are knowledgeable of the subject matter and who can legitimate that assertion as knowledge (see Prelli, 1989, p. 112). Despite the presence of the term "technical ornithology," there was no such audience of specialists in late nineteenth-century American ornithology, except for a few scientists like Ridgway, Coues, Allen, and Chapman, whose museum work kept them in regular contact with ornithology. In fact, the early polemics by authors like Allen and Brewster excerpted in chapter 5 were attempts to establish the writers themselves as such specialists. Polemics like these are typical of emerging professions (see, for instance, Lankford, 1981), though biologist Stephen Jay Gould (1993) considered such "opinion pieces" a fully accepted, if infrequently practiced, genre of contemporary science writing as well (p. 321). Still, most of the writers of *The Auk* during this first phase of ornithological discourse were members of other fields—doctors, lawyers, bankers—and so did not have any shared authority to authorize or cancel any writer's claims. Thus, the mere union of bird enthusiasts sharing common interests in ornithological study, and even the publication of their findings in a journal, did not automatically transform either the union or its discourse into professional entities. Henrika Kuklick (1980), for instance, found that the field of sociology had similar experiences.

In addition, early American ornithology had only one major research topic, the taxonomic classification of American birds, and only one major methodology, the individual components of which included observation, collection, and description, equally accessible to both amateur and technical ornithologist. But this taxonomic focus did not assure ornithology of professional status. As Lawrence Prelli (1989) noted, "Taxonomies have persuasive power when their categories are assumed to be individually discrete and collectively exhaustive" (p. 194). Late nineteenth-century ornithology, on the other hand, was striving to complete its taxonomy of American birds; so the taxonomic-power *topos* would not have been in place as a problem-solving topic. Thus, the incompleteness of its taxonomic *topos*, the absence of other scientific *topoi*, its loosely bound membership, and flourishing amateur involvement indicate that late nineteenth-century American ornithology was not yet a legitimate science.

Shifting Discourse in the Early Twentieth Century

A transformation, however, did occur within American ornithology during the early twentieth century, as the emergence of natural history discourse during the field's second discourse period has shown. This change can be attributed to at least two factors, one internal to American ornithology and the other an external historical event. After a quarter century of intensive field and laboratory studies by amateur and professional, taxonomic issues had become less challenging, causing ornithologists to search for other lines of research. Further, with the rise of conservation and preservation movements, bird enthusiasts, particularly amateurs, could direct their attention to other organizations besides the American Ornithologists' Union. These factors altered two very basic characteristics of American ornithology—its research and its membership.

Taxonomic issues in early American ornithology

John Ziman (1984) suggested that "scientific theories appear as ordering principles that explain general classes of observational and experimental facts, including the *taxonomies*, 'laws,' causal chains and other empirical regularities that are discovered about such facts" (p. 28, italics added). Ornithology's early concern with taxonomic classifications of birds was its strongest connection with empirical science, a connection traceable from the seventeenth-century studies and discourse of John Ray and Francis Willughby through the trinomial classification schema for American birds developed by nineteenth-century American ornithologists. Taxonomic studies proliferated both among technical ornithologists working in museum laboratories during the late nineteenth and early twentieth centuries and among amateurs studying bird skins at nearby museums, both groups subsequently publishing their comparative studies in *The Auk*. The journal has continued to publish both field and laboratory studies suggesting taxonomic changes, as well as more recent DNA research to reclassify birds genetically. This early and sustained emphasis on taxonomy helped to associate even early ornithological study with experimental science, which focuses on legitimate scientific theories. The preoccupation during early American ornithology with taxonomic questions by both amateur *and* technical ornithologist was a likely cause for their continued association in the same discourse community, because they possessed common theories, *topoi* (that is, "taxonomy"), and a shared image of what it meant to be an "American ornithologist."

It may be argued that taxonomy not only had become an established *topos*, but also had generated a body of associated theory. However, American ornithology had fairly well exhausted the classification of American birds by the early twentieth century, at least to the extent that contemporary methods, instrumentation, and theories allowed. There were, of course, still other species to be identified and more exact range limits to be mapped; but, given the state of knowledge at the time, the general classification of American birds had become clear. Thus, American ornithology began seeking other *topoi* toward which to direct its attention. The void left by the emptying of the taxonomic *topos* is a probable cause of the dramatic changes seen during this second period of American ornithological discourse history. At that time, natural history writing and its associated methodology, which is more accessible to the amateur, surged to the forefront of ornithological discourse, filling the void left by the decline of taxonomic study. Since the roots of technical ornithology were set firmly in taxonomic study, the disappearance of taxonomy as the central *topos* of American ornithological study left the field without clear focus. Thus, progress toward the professionalization of American ornithological science in the second and third decades of the twentieth century was in a sense set back by the decline of taxonomic questions. As a result, interest turned to behavioral and ethological topics and methodology, which were observationally based and therefore accessible to the amateur ornithologist. This changing interest accounts in part for the rise of natural history discourse during this period of American ornithological history, as researchers began to pay more attention to the behavior of birds and subsequently developed methodology to quantify behavior. Thus, the emphasis upon observation-based research and the continued publication of personal observation notes and field checklists created the conditions for the continued rise of the personal narrative, to which both amateur and technical ornithologist could contribute.

This major shift in American ornithology, however, was not a Kuhnian revolution resulting from a shifting paradigm (Kuhn, 1970). In the first place, American ornithology was yet to be a science with developed paradigms. More accurately, the taxonomic paradigm, if there was one, had not lost its explanatory power; it had just lost its ability to sustain ornithologists' research. Thus, these events more accurately reflect the view proposed by Lakatos (1978) that science operates through viable research programs, because, for ornithologists, taxonomy's research program had simply dried up. Without doubt, there were more birds to be discovered; but American ornithologists believed that their present taxonomic schema would absorb these new discoveries. For ornithological research to continue, new topics were sought, new methodologies explored.

Historical obstacles to a merging of amateur and technical worldviews

Drawing on Kenneth Burke, Prelli (1989) observed that different terministic orientations may never achieve lasting identification (pp. 20–21). In other words, the worldviews of amateur and professional may be so disparate that the two may not easily form one unified community. Such diverse orientations as those generating studies in natural history, which focuses on living things, and experimental science, which views all things as objects for study and manipulation, are in direct opposition to each other. The opposition inherent in these worldviews was an underlying problem, which would eventually disrupt the collegiality among early amateur and technical ornithologists.

The establishment of museums through the end of the nineteenth century and the increasing interest by European scientists in obtaining American bird specimens for classification purposes created a great demand for bird skins. Such specimen collection was an activity staunchly defended by technical ornithologists, an attitude strikingly evident from the "more than a score" of ornithologists of the 1880s with private bird collections, some having 20,000 to 60,000 specimens *each* (Mayr, 1975, p. 367)! Many amateurs, however, did not favor such activity. Disparaging remarks about specimen collecting from natural history writers such as Thoreau and Burroughs are exemplary. There even emerged a technical ornithologist, Ludlow Griscom, who advocated an alternative approach to bird study. Griscom, an ornithologist at the American Museum of Natural History in the early twentieth century, showed that the use of the field binocular along with careful observation of field marks, behavior, and song could just as accurately identify birds in the field. The practice of collection was further strained by the controversial slaughter of birds for the millinery trade, a market that eventually dried up with the advent of the conservation movement at the turn of the century.

Accompanying the protest against feather exporters was the spin-off of a specialized ornithological discourse, a discourse in support of the conservation of birds. The main voice of this movement was the National Audubon Society's *Bird-Lore*, begun in 1899 by ornithologist Frank M. Chapman. This split of ornithological discourse into technical and conservation emphases began the divergence of ornithology into professional and amateur camps, concrete evidence of the professionalization of science by specialization and exclusion.

Killingsworth and Palmer's study (1992) of ecologists' writing revealed another force fueling the break between amateur and technical ornithologist: environmentalism. Just as leaders like Theodore Roosevelt and Gifford Pinchot influenced the creation of government agencies like the Department of the Interior and university departments like forestry and wildlife management in support of conservation, John Muir and the Sierra Club were advocating a wilderness ethic in support of a more radical preservation of the wild (pp. 23–25). Environmentalists' call for a return to nature and for the preservation of the natural world was an ethic with which amateurs, focused on nature and the natural history world as well, could easily identify.

Thus, bird aficionados now had two other movements—conservation and preservation—occupying their attention. As a result, American ornithology had begun to lose some of its amateur participation at the turn of the century as the National Audubon Society and Sierra Club began attracting more and more amateurs. In effect, these two movements served as filters in the slow evolution toward professionalization of American ornithology by initiating the process of exclusion.

The 1920s and 1930s: A Defining Period for American Ornithology

What eventually rerouted American ornithology back on its road to professionalization was something analogous to inter-paradigmatic science, which is "characterized by the absence of any single dominant paradigm commanding consensus among researchers" (Prelli, 1989, p. 93). For American ornithology, its cooperation with other scientific fields such as ethology, anatomy, and physiology provided theory, methodology, and *topoi* that would eventually give the emerging science a clear focus for the future. Ornithologist-historians consider the 1920s a defining period for the science, because zoologists began to realize both the interconnection among disciplines and the value of birds in the search for causation (Sibley, 1955, p. 630). At this time, animal behavioral studies began to emerge, particularly through the influence of Europeans like Konrad Lorenz in the 1930s. A perusal of the early volumes of *The Auk* shows that ornithologists had been studying birds in this manner all along. Birds were not only easily observable subjects; there was also such a great diversity of them that ornithologists had many years of readily available research topics. In addition, the principles of Darwinian evolution were now firmly in place. Science recognized the protracted process whereby species interact with their environment and one species influences another. Thus, through the example of Darwin's finches and bolstered by ornithologists' already established taxonomic research, American ornithologists could legitimately expect to contribute to the growing volume of scientific research in evolutionary and animal behavioral studies. European ornithologist Erwin Stresemann (1951/1975) has asserted that interchanges among scientific fields were responsible for ornithology's advancement in Europe (pp. 351–364). The same can be said for ornithology in the United States between 1920 and 1940.

As chapter 5 has shown, the year 1930 proved to be an especially significant one in the development of American ornithological discourse, which was quickly becoming more diverse as American ornithology began to locate itself within scientific paradigms. Ornithologists showed that the study of birds could improve the agricultural economy and were beginning to show how birds' welfare could be attached to ecological health, a topic that would be of major concern in only a few decades. *The Auk* was losing its provincial air by publishing bird research from other parts of the world. There were also attempts to standardize methodology. Most importantly, increasing empiricism was spurred on by the beginnings of instrumentation, a crucial development in the professionalization of a science (Latour, 1987; see Lankford, 1981) and the foreshadowing of the technological advances to be brought about by the rise of the industrial-military complex after World War II. This diversity followed shortly after the expansion of American ornithologists' research topics during the 1920s as a result of new interconnections between ornithology and other sciences. The expanding vision would bring the discipline to the point of professionalization during the next period.

Period of Interplay (1940–1970): The Critical Period for American Ornithological Discourse

In 1965, the Research Committee of the American Ornithologists' Union conducted a survey of current ornithological research projects in the United States and Canada. The results show the extent to which American ornithology had made interconnections with other sciences by mid century (Notes, 1965, p. 320). In the following list, note especially the large number of projects in ecology, physiology, and economic ornithology, as well as the waning interest in birds' life histories and studies of plumage, once common topics in *The Auk*:

- behavior - 65
- ecology - 59
- physiology - 51
- systematics & distribution - 42
- migration - 38
- economic ornithology - 30
- regional studies - 28
- anatomy - 18
- evolution & genetics - 17
- life histories - 14
- parasites & diseases - 9
- plumage - 8
- paleontology - 2

Historically, a number of events signal the changing face of American ornithological science during this 30-year period. Central to professionalization in science is the recognition of the prominent role played by scientific research and the need to assist in its funding. In 1947, two funds were established to support ornithological research. The Louis Agassiz Fuertes Research Grant aided aspiring ornithologists, and the Frank M. Chapman Memorial Fund was founded in order to provide grants and fellowships for ornithologists. As evidence of its importance, by 1962 awards by the Chapman Fund had risen to $84,000 (Notes, 1963, p. 225). In 1958, ornithology was singled out as one of the fields of interest in a new post-doctoral fellowship in biological sciences at Yale (Notes, 1958, p. 118). In 1960, the AOU established its own Van Tyne Research Award in order to provide yearly assistance in the funding of ornithological research. Also in this same year, the AOU began publishing as a separate article the yearly reports of the AOU's Research Committee in an attempt to encourage and promote an exchange of ideas for ornithological research (Report, 1960, p. 78). In contrast to the prior period when *The Auk* contained only scattered announcements of research awards, by 1966 issues of *The Auk* regularly contained two to three full pages devoted to announcements of research grants, fellowships, and other research-related topics.

Other events of the period 1940–1970 indicate a rapidly expanding volume of ornithological research for American ornithologists to absorb. Perhaps an early indication is the appearance in the April 1955 *Auk* of a book review section summarizing and critiquing key published works of interest to American ornithologists. In 1959, the AOU Research Committee began discussing the possibility of centralizing into one publication current literature reviews on ornithology, because researchers were finding it increasingly difficult to conduct their literature reviews since abstracts of ornithological papers were scattered among many journals (Report, 1960, p. 78). In 1961, the AOU began negotiations with *Biological Abstracts* for the distribution to all AOU members of its yearly abstracts dealing with birds (Report, 1962, p. 263). In this same year, the Research Committee's discussion of centralization had expanded to include the computerization of raw data for future use by ornithological researchers (p. 263). And perhaps most important of all, in 1954 job placement services for ornithologists had become a consideration in the AOU's decision to join the American Institute of Biological Sciences, which offered these services (Notes, 1954, p. 494).

A number of discursive, linguistic, and historical changes described in chapter 5 also signal this third period as critical in the development of ornithological discourse. In essence, these changes signal the professionalization of American ornithological science. By the end of the period, laboratory articles had become the most numerous kind of article, describing complicated experimental techniques rather than simple specimen studies. In particular, studies of avian physiology showed ornithology's continued merger with other scientific fields. Most notable were references to sophisticated instrumentation and empirical methodology to buttress scientific claims. The increasing use of Acknowledgments sections points to an increasing sense of community. Trends regarding the quantity and length of articles completely reversed themselves after 1960. The use of graphics, particularly graphs and tables, dramatically increased after 1960 as well. Linguistically, nominalizations and technical terminology began their rapid rise during the period—a particularly significant change since discourse is not truly specialized without exclusion, reflected here in increasing technicality. The trend in the use of markers for reliability of evidence likewise reversed itself, indicating changes in argumentation within the discourse. And by the end of the period, most contributors to *The Auk* were professionally affiliated, an increase paralleling the increasing empiricism and experimental research articles in the journal. The decline of amateur contributions as symbolized by the quickly disappearing addresses of unaffiliated authors during this period was another sign that American ornithology had specialized and become a more tightly knit, exclusive organization. As Prelli (1989) commented, a typical strategy for making an individual's relationship with a recognized scientific community ambiguous is to note that person's failure to publish in sanctioned scientific journals (p. 130). O'Connor and Meadows (1976) have also reported the importance of the professional journal in the professionalization of British geology. Influencing this increasing specialization are two historical events: (a) the establishment of a university program for ornithological study, and (b) the rise of environmentalism, which impacted ornithology's membership.

University curricula as a sign of professionalization

One explanation for the changing face of American ornithology and its discourse during the period 1940 to 1970 can be traced to the development of paradigms controlling scientific thought and research (Kuhn, 1970). These paradigms must likewise be transmitted to aspiring new members of the profession; otherwise, the scientific community would not be able to perpetuate itself, and in ornithology's case would not have become a profession at all (see Kohler, 1982; O'Connor & Meadows, 1976, p. 79).

Consequently, American ornithology needed an established educational program to introduce would-be ornithologists into its paradigms, philosophy, and sense of community. As Prelli (1989) pointed out, "Scientific training is aimed at teaching students to think and to act as scientists, according to the orthodox beliefs, problem-solutions, and procedures of some substantive research area" (p. 91) (see also Weimer, 1977).

In the late nineteenth century, American ornithology as a newly formed organization had no such educational program. American universities, though training students well in zoology and botany, did not have any established programs of study in ornithology. Rather, the study of birds was centered in the system of state and national museums in the United States. Even though their specimen cabinets were relatively open to any ornithologist, such specimen studies were isolated affairs between bird student and bird skin. Frank Chapman (1935), for instance, left his banking job to "become an ornithologist" with little formal scientific training. He acquired his wealth of knowledge from his exuberant interest in the subject, hands-on experience with hundreds of study-skin collections, and frequent collection trips around both American hemispheres. As he acknowledged, he was introduced into the community of ornithologists through his attendance at yearly meetings of the American Ornithologists' Union, associations with other museum curators, and personal contacts and correspondence during his work at the American Museum of Natural History in New York. Thus, rather than having attended a formal college, Chapman's education could be said to have been obtained from what Diana Crane (1972) has described as an "invisible college," a "communication network" (p. 35) of scientific audiences interconnected through similar research interests.

Chapman's case was not unique during the latter part of the nineteenth century, for the AOU's ranks contained members from many walks of life, in essence ornithological moonlighters. Numerous early American ornithologists were practicing physicians, deriving their interest in and knowledge about birds from zoology and medicine as, for instance, Spenser Fullerton Baird, late-nineteenth-century Smithsonian curator and organizer of the federal government's specimen-collection trips through the American West. Coming from such diverse backgrounds, American ornithologists in the early history of the science had no institution to derive an understanding of research goals, standards, or values—that is, those scientific paradigms unique to American ornithology. At this time there were a few generalized handbooks, notably those by Elliott Coues. However, in lieu of standard textbooks, the pages of *The Auk* contained the only compendium of written principles guiding the "science" of American ornithology. But, because the journal itself was interwoven with amateur natural history discourse, it was unable to provide the needed transfer of paradigms, which in late nineteenth-century American ornithology were only in the formative stages. How then was a reader of *The Auk* able to distinguish amateur from technical ornithology?

The absence of formal ornithological training would soon change, however, with the arrival on the ornithological scene of Arthur A. Allen. As assistant professor of ornithology, he began his tenure at Cornell in 1916 and soon built a program of ornithological study that rivaled any in the country. His special expertise in life histories of birds and in environmental and behavioral studies put him on the cutting edge of expanding ornithological research in the 1920s; and he quickly broadened his program to include new technologies, such as still and motion-picture photography and sound recording. Through Allen's leadership, Cornell was the first university to offer a doctorate in ornithology, and his "Grad Lab" grew in reputation during the 1930s. His program at Cornell was a goal of any serious young scientist aspiring to become an ornithologist between the 1920s and the early 1940s. His *Ornithology Laboratory Notebook* (1927) had five editions, and his text *The Book of Bird Life* (1930) was reprinted 11 times (Pettingill, 1968, p. 193). By his retirement in 1953, he had taught over 10,000 undergraduates and had guided over 100 graduate students through advanced degrees in ornithology (p. 200), including some of the most prominent future leaders of American ornithology. Thus, by the 1950s, numerous programs of ornithological science in American universities were now actively disseminating ornithological training and its accompanying paradigms. And because many of these founders had received their own training largely from one source, Arthur Allen, they would have brought with them a common understanding of "American ornithology" and its developing paradigms. Such educational training was also the means whereby the community's research models and shared values were passed on. This is most evident in the comments of Allen's obituarist and former student of the early 1930s, Olin S. Pettingill (1968):

> All of us in the Grad Lab greatly valued the benefits from working in close association: the mutual stimulation derived from common interests, the discussions, the sharing of concern with our respective problems, and the consequent attainment of new concepts that neither classes nor texts could provide. The end result was a sense of fraternity and a feeling of pride in accomplishment. Ornithology at Cornell was something special and we were helping to make it so. (p. 195)

This Allen/Cornell web of information and training was not only a means for disseminating ornithological paradigms. A sense of community would have been instilled in students by their associations with ornithologists and classmates in their respective departments. In addition, attendance at up to three yearly ornithological meetings, those of the national AOU, the East's Wilson Ornithological Society, and the West's Cooper's Ornithological Club, would have enabled young ornithologists to form acquaintances with associates from other universities as well. Such professional associations provided attendees a sense of unity, of being members of an established society.

Clearly, by this third period American ornithologists saw the need to take an active role in encouraging these future ornithologists. In 1947, the newly established Fuertes Research Grant was targeted specifically to young aspiring ornithologists. In 1952, the AOU began awarding free student memberships, the first year to a group of 20 young ornithologists. By the 1960s, the yearly award had more than doubled. Also by this time, the AOU customarily offered financial assistance for "young ornithologists" (Notes, 1960, p. 115) to attend its annual meetings; and in 1962 the organization extended its assistance so that some aspiring members could attend the International Ornithological Congress being held in Ithaca, New York (Notes, 1962, p. 136). Even in the awarding of the Van Tyne Research Award, the novice was given special consideration: Where projects were of equal importance, the founding AOU Research Committee agreed that preference be given to the "young investigator . . . and to investigators without good opportunities for securing funds elsewhere" (Report, 1961, p. 243). Thus, the results of the now established educational training of American ornithologists would have come to fruition during the period of discourse interplay. The common understanding of what "American ornithologist" meant gave impetus to the professionalization of American ornithology, which, as this study has shown, began in earnest during this period.

Movement Toward environmentalism

Also at this time, the forces of environmentalism widened the gap between those advocating the preservation of nature, and science, technology, and business advocating progress and profit. Killingsworth & Palmer (1992) noted that the transcendentalist language of wilderness preservation was on the rise after World War II as the environmental movement took hold (pp. 23–25). In the process, several scientists attempted to transform scientific discourse into public rhetoric by transferring scientific objectivism into ecological awareness. For instance, in his attempt to sidetrack what he considered needless development, Aldo Leopold spoke of the land in transcendental terms in order to advocate its preservation for scientific, recreational, and wildlife purposes:

> Land, then, is not merely soil; it is a fountain of energy flowing through a circuit of soils, plants, and animals. Food chains and the living channels which conduct energy upward; death and decay return it to the soil. The circuit is not closed; some energy is dissipated in decay, some is added by absorption from the air, some is stored in soils, peats, and long-lived forests; but it is a sustained circuit, like a slowly augmented revolving fund of life. (1949/1987, p. 216)

In 1962, biologist Rachel Carson's best-selling *Silent Spring* brought to the public's attention the dangers of chemical toxins through her apocalyptic narrative and logical, convincing presentation of scientific data. Her rhetoric was aimed not only at developers, but also at scientists, whose research had laid the foundation for the poisoning of the environment. Also at this time, university biologist Barry Commoner criticized the cooperation of science with business and technology, thereby implicating science as an accomplice in the environmental crisis (Killingsworth & Palmer, 1992, pp. 53–93). For American ornithological discourse, the significance of public interest in the environment at mid century is that natural history discourse also had its origins in transcendentalist themes from writers like Henry David Thoreau and John Muir. Thus, those amateurs attuned to natural history themes would relate readily to the concerns of environmentalists, particularly since their discourse functioned in the natural history world. Amateurs questioning the ultimate value of scientific research in improving life would be directed away from science, including ornithology, as a means of satisfying their desire for nature study. The ultimate result was that those with personal rather than professional interest in nature, including birds, were directed toward environmental organizations and away from scientific associations, further isolating the amateur from the professional scientist, and vice versa. The rift was exacerbated by scientists' insistence on objectivity and distance from human affairs. Since environmentalists had as their aim social, ethical, or political action, scientists and their discourse remained aloof.

Though there are many reasons for changes in an organization's membership figures, the decided decline in membership in the American Ornithologists' Union from the mid-1950s to the early 1960s, seen particularly in light of the extremely sharp increase during the early 1950s, perhaps reflects the changing nature of the membership, that is, the departure of the amateur (see Figure 6.1). In the very least, the graph is further evidence of conflicting forces at work during this period of discourse interplay.

Further widening the rift between scientist and non-scientist was public education, which, as Taylor (1996) pointed out, has become essential to contemporary science as a means through which it attracts and socializes new members (p. 161). An examination by Killingsworth and Palmer (1992) of science

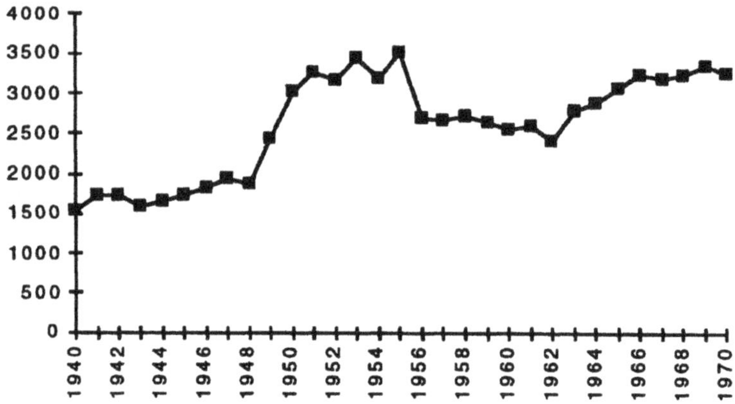

FIGURE 6.1. Membership totals for the American Ornithologists' Union from 1940 through 1970.

text-books showed that school children of the 1960s were being indoctrinated into liberal environmentalist perspectives (pp. 93–99). This influence can be viewed in contrast to Houghton Mifflin's widespread introduction of John Burroughs nature clubs into schools in the early part of the twentieth century. Young nature enthusiasts skimming the pages of *The Auk* of the early 1900s would find much familiar to them after reading Burroughs' descriptive personal accounts of birds, whereas young mid-century readers had no such association with *The Auk* of the 1960s. As Taylor commented, "Science, as a cultural discourse, can be read in many, even contradictory, ways. In this sense, the image of science, as defined by particular demarcation strategies, is a product of both historical evolution and rhetorical negotiation" (p. 109). Aspiring scientists of the 1960s would soon lose their environmental perspective, thus distancing even further the professional-to-be from the young amateur, as Killingsworth and Palmer noted:

> ... our research suggests that environmentalist values have very little influence on mainstream scientific research and that all such ethical systems are thoroughly repressed in late undergraduate and graduate education, during which time the conventions of the research discourse are learned. (p. 99)

It was within this post–World War II milieu that American ornithology was establishing its own unique identity.

The decline of natural history discourse

As described in chapter 5, natural history discourse and the personal narrative still persisted on almost equal footing with experimental science discourse during this transitional period to professionalization. In his study of biologists' writing, Myers (1990) suggested that writers and readers of professional and popular science discourse should not be able to "enter easily" into each other's discourse (p. 143). The fact that these two discourses were easily interchangeable during much of this third period of American ornithological discourse is another indication that professionalization was still in process.

One reason that ornithology has attracted the amateur over the years is its continuing popularization, even by the science itself. For instance, Arthur Allen followed Frank Chapman's lead by publishing many popular articles about birds in *National Geographic* and *Bird-Lore* magazines, as well as two volumes of natural history accounts for school children in the 1930s. He was likewise a lifelong public lecturer on birds.

Another reason for the continued intermingling of the two discourses was the observation-based research American ornithologists were doing—a situation also occurring in British geology during the twentieth century (O'Connor & Meadows, 1976). In particular, ornithologists' behavioral and life history studies required them to report what they had observed, now on a more technical level, but still of necessity using the same inventory of reporting verbs (that is, *see*, *hear*, *observe*, *record*, etc.) and the same range of behavioral action verbs (that is, *perch*, *roost*, *fly*, *ate*, etc.) as earlier writers used. In other words, the language itself permitted the continued interplay of discourses. But as ornithologists refined their behavioral observations by adding terminology to describe specific actions (for example, *feather settling*, *sally gleaning*, *sweeping*, *freezing*, *appeasement*, *threat*, *alert posturing*, etc.), researchers now had a growing vocabulary, which had to be mastered before attempting field observation. This specialization is reflected both in the decreasing numbers of unaffiliated authors during the interplay period and in the predominance of experimental science over natural history articles in 1960, after which natural history discourse would continue to decline.

Let us now reconsider the questions posed in chapters 3 and 4 for identifying the discourse models operating in American ornithology. Ornithologists' answers to these questions are radically different by the end of this third period of American ornithological discourse, a short span of 70 years:

- Who am I? Not just an interested observer of birds who dabbles in ornithology as a pastime, but a university-trained scientist who expects to make contributions to accumulated scientific knowledge and to a better quality of life.
- For whom am I speaking? Not just for peer bird enthusiasts and technicians who want to know where new birds can be located, but for a community of scientists with a shared worldview.

- What forces bind members together? Not just a yearly AOU meeting, occasional correspondence, and *The Auk*, but a university education, a growing volume of "literature" from many fields, multiple regional and national meetings, and a diversified research agenda.
- What is their worldview? Not just to contribute "something" to advance the knowledge of American bird life for anyone who wants to listen, but to clarify and, with luck, change scientific paradigms affecting ornithology for those knowledgeable enough to recognize and assent to the change.
- What is the status of their knowledge? Not just based on direct observation and personal reading, but informed by a set of developing paradigms acquired through a rigorous, formal program of advanced study.
- How are knowledge claims made? Not just by saying so, but by demonstrating a recognized methodology and by placing one's claims within the ongoing body of research and before an acknowledged forum of one's peers.
- How is the same knowledge used by both groups? Not just to locate birds, but for the amateur to increase understanding and appreciation for birds and for the scientist to obtain professional recognition and advancement.

Revised Discourse Schema

Figure 6.2 illustrates the forces at work in shaping American ornithological discourse. As chapter 5's analysis of *The Auk* has shown, there has been a gradual progression throughout American ornithological history toward experimental science discourse, as seen by the arrow on the Continuum of Scientific Perspectives pointing toward experimental science topics and methodology. The shaded area represents the period of transition, when discourse change has occurred from the rise of natural history discourse in the early twentieth century until the predominance of experimental science discourse after 1970. The prominence of natural history discourse in the early part of the century corresponds to the greatest disparity between the two poles of discourse, thus the largest amount of discourse variation and, consequently, the longest vertical line separating the two discourse models. As the century progressed and experimental science discourse slowly but gradually dominated, indicating an end to discourse instability, the shaded area likewise disappears. However, in the process there is an interplay of the two discourses between 1940 and 1970 represented by the jagged lines, as social, historical, linguistic, and scientific forces react and interact to influence increasing professionalization. The point of intersection between discourse variation and discourse continuum symbolizes writers of transitional discourse, like ornithologist Frank M. Chapman

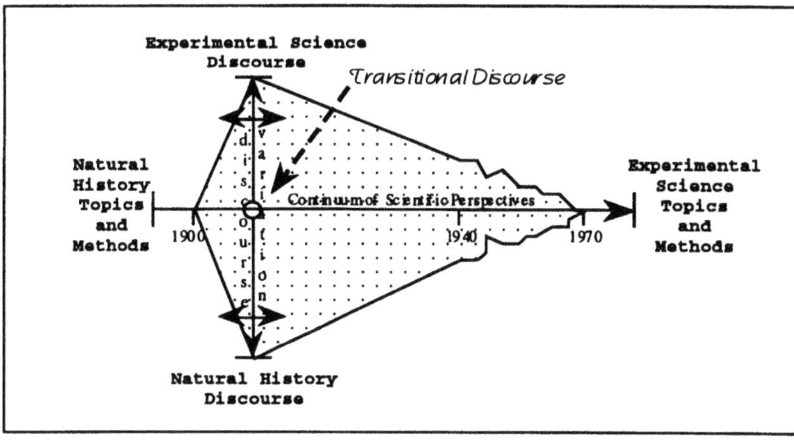

FIGURE 6.2. Schema representing the evolution of American ornithological discourse.

described in chapter 4, who function equally well in both types of discourse. Quantifiable changes in their writing format and style can be considered benchmarks in the transition from one discourse pole to another. With the professionalization of American ornithology after 1970, the discourse is again stable because the field is now a science dominated by the professional ornithologist.

REFINING THE DISCOURSE MODELS

It is now possible to merge the findings of chapters 3 through 5 in order to gain further insight into the models used during this study. First, the three discourse models proposed earlier can now be refined in light of the writing practices observed in *The Auk*. In addition, an evaluation can be made of the usefulness of the Continuum of Scientific Perspectives on Nature in describing the science of American ornithology.

The Discourse Models

The analysis of articles from *The Auk* in chapter 5 permits a reevaluation of the discourse models in chapters 3 and 4 in light of American ornithological discourse. The linguistic features proposed for both the natural history and experimental science discourse models stand up well against the findings in chapter 5. Though the experimental science model predicts the disappearance of personal pronouns and an increase in causal clauses,

American ornithologists continued to use personal pronouns, though less frequently than before; and there was no evidence of increasing use of causal clauses. Because these characteristics were likely caused by the unique place of natural history in American ornithology in the first instance and, in the second, by the minimal appearance of theoretical argumentation in the relatively young science of American ornithology, the evidence does not suggest the need to revise the standard discourse model as generally representative of science.

However, the transition model does warrant a review. Because American ornithology employs a transitional, developing discourse, its characteristic features should be incorporated into the transition model. Changes suggested by ornithological discourse are that

- average sentence length decreases rather than increases
- the number of simple sentences increases rather than decreases
- the percentage of interrupting clauses significantly increases
- causal clauses do not increase

Also, five features posited by chapter 4's transition model as changing early in the transition process were actually late in registering changes in American ornithologists' discourse:

- the appearance of technical and polynomial terms
- less emphasis on subjects as place-holders
- decreased use of the verb *to be*
- an increase in the number of simple sentences
- emphasis on reliability of evidence

One additional feature, the appearance of citations, showed marked change early rather than late in contrast to the transition model.

Adjusting chapter 4's transitional discourse model in light of these changes produces the revised lists below, which suggest significant linguistic changes during the transition from natural history to experimental science discourse. It is proposed here that these discourse features may be considered benchmarks, or *discourse indices*, signaling the professionalization of a science. Those asterisked, in combination with increases in the use of high-density graphics like tables and graphs, have the greatest predictive value as indices of professionalization because changes in these indices were the most rapid and clearly evident in American ornithological discourse. These 15 features, along with increased use of high-density graphics, should be valuable indicators for other rhetorical and linguistic researchers studying emerging sciences:

Early Evidence of Change (the beginnings of professionalization)

1. average sentence length decreases
2. * nominalizations increase
3. * the percentage of subject-verb interruptions significantly increases
4. * increased use of citations

Late Changes in the Transition Period (indicative of professionalization)

1. increased use of metadiscourse
2. minimal use of personal pronouns
3. increased use of definitions
4. * increased use of technical and polynomial technical terms
5. less emphasis on subjects as place-holders
6. decreased use of the verb *to be*
7. * increased use of simple sentences
8. an emphasis on reliability of evidence over other markers of evidentiality
9. preference for passive voice
10. restricted use of complex clauses
11. restricted use of evidentials

A Revised Continuum of Scientific Perspectives

The data used in this study have been gathered from two different databases from *The Auk* and have measured multiple features, for instance, linguistic, graphic, and formatting changes. Despite their varied origin, the data mesh nicely with Killingsworth & Palmer's Continuum (1992) and reflect the changes one would expect, given the Continuum's assertions. For American ornithology in particular, the analysis of *The Auk* has shown that American ornithology has likewise passed through the stages outlined by the Continuum, with one exception: A category between "observing in the field" and "observing and measuring in the field" should be added to reflect scientific specimen collection as the second stage in American ornithological development (see Figure 6.3). This revised continuum would then be a more useful guide in the study of other scientific fields that have employed specimen collecting as an early methodology. Though some painters and natural historians like Audubon were also specimen collectors, the technical ornithologists, rather than the natural historians, used the collection of specimens as their primary field methodology for advancing their taxonomic work in the late nineteenth century. "Observing and measuring in the field" did not begin in earnest until the early twentieth century, when bird banding was incorporated into ornithological field

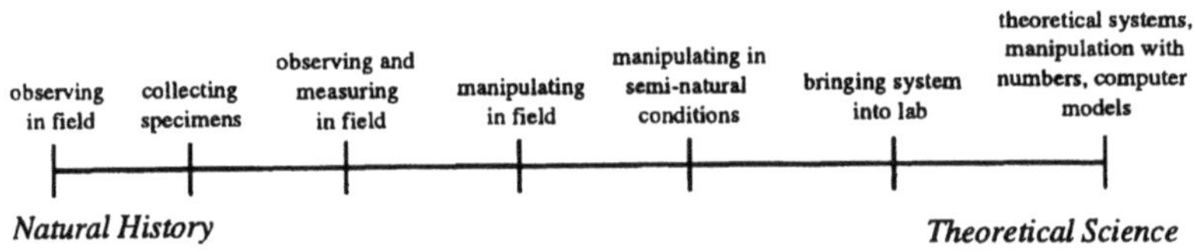

FIGURE 6.3. Revised Continuum of Scientific Perspectives on Nature. Adapted from Killingsworth & Palmer, 1992, p. 112, with the permission of Southern Illinois University Press.

methodology. Thus, specimen collection was an intermediary stage of methodological advancement and an early means of separating the amateur from the professional ornithologist.

ORNITHOLOGY AND SCIENTIFIC DISCOURSE

Merging the findings of this study of American ornithological discourse with those by Bazerman (1988), Abt (1981), the Physics Survey Committee (1973), and Killingsworth and Palmer (1992), it is apparent that many generalizations about scientific discourse implicit from previous studies do not hold for American ornithological discourse and may be overgeneralizations about scientific discourse as a whole. Thus, there is a need to reevaluate the forces that interact in the creation of professional scientific communication.

Because the evolution of American ornithological discourse has had its own unique history and characteristic features, a study of ornithological discourse sheds further light into the development of scientific discourse and consequently into the factors that characterize the writing of scientists. More importantly, an analysis of ornithological discourse via *The Auk* is distinguished from previous journal-based studies of scientific discourse in that this study shows the progression of a science, American ornithology, from its inception, whereas studies like Bazerman's and Abt's are entering the discourse of physics and astronomy *in media res*. As the last two chapters of this study have shown, American ornithology has experienced the continuing participation of amateur contributors, as well as the persistence of methodology such as bird listing, counting, and behavioral observations, all of which permit a protracted association between the amateur and professional ornithologist. The influence of the nonprofessional in ornithology has delayed the specialization of its discourse and has supported previous studies and theoretical arguments of philosophers of science as to the essential role played by social forces in a developing scientific discourse. Sophistication in both methodology and theoretical articulation during the mid twentieth century precipitated the rift between natural history (or amateur-originated) discourse and experimental science (or professionally generated) discourse. It was not until this time that American ornithological discourse clearly demonstrated the discourse features described by previous researchers as characteristic features of scientific discourse.

It is at this point, then, that we can say from the perspective of its discourse that American ornithology has professionalized:

- Its discourse is now characteristic of an experimental science rather than a natural history model, signaling a transfer of power from amateur naturalist to trained scientist.
- It now focuses on more theoretically complex issues.
- It is now packaged in a standardized format that provides for the presentation of information traditionally considered of "scientific" concern (Gross, 1990, pp. 85–96).
- Argumentation is supported with theoretically complex graphical displays and scientifically accepted statistical inferences.
- From the viewpoint of philosophers of science, the social forces necessary to categorize a science as professional are now evident in American ornithological discourse.
- Well-established Acknowledgments sections and an interconnecting web of citations demonstrate the presence of an association of professional peers.
- Its technical terminology and methodology have become refined enough to limit the audience that can fully comprehend it.

Its discourse, then, unequivocally defines American ornithology as professional. Some might assert that the founding of the American Ornithologists' Union in 1883 or the establishment of university degree programs in ornithology in the 1930s signaled the professionalization of American ornithology. However, as seen by the priority given to the written scientific record even as far back as the seventeenth century (Gross, 1990, pp. 174–176), written discourse is at the center of the transmission of scientific knowledge. Thus, science cannot be said to have truly professionalized until its discourse has bridged the gap from amateur to professional.

Changes in Article Length and Quantity

Previous analyses of astronomers' and physicists' discourse have shown the increasing length of science articles during the twentieth century. Abt (1981), for instance, cited the years 1940 to 1980 as the period of greatest increase in the length of astronomers' papers. Bazerman (1988) connected the increasing length of physicists' discourse with "increasing knowledge and theoretical elaboration" (p. 164), requiring added background and context. Though the length of ornithologists' articles in *The Auk* also increased in length during its first 70 years, their articles, unlike astronomers' and physicists' discourse, have *decreased* in length since 1960. Since chapter 5 has shown that ornithology has also increased in knowledge and theoretical elaboration, especially in the last 20 years, assertions associating discourse length with epistemological and theoretical sophistication may at first seem negated here. However, this seeming contradiction in the data between physics and ornithology can be explained, at least in part, by the influence of the amateur, or novice, writer in ornithological discourse and consequently by the changing forces of professionalism in American ornithology. An amateur science, for instance, will not have an elaborate web of technical terminology with which to argue and develop theoretical stances. Nor will it have developed an agreed-upon methodology to which its members adhere. Nor will rhetorical or theoretical topics be established in a developing science. The writers of such discourse, as in American ornithology in its first 70 years, must compensate for the absence of an accepted terminology, methodology, argumentation, and *topoi*, the shared meanings of which enable users of the discourse to communicate efficiently. In the absence of accepted theoretical paradigms, writers must elaborate more fully upon the topics they present; in the absence of accepted technical terminology, they must define more fully the words they use; and in the absence of accepted methodology, they must describe more completely the methods they select—thus, justification for an increasing article length. Bazerman's study confirmed fuller descriptions of methodology early in a developing field, because his research revealed a proportional decline of methods and apparatus sections in spectroscopy articles over the years (p. 175). Already noted also is the argumentative shorthand provided by statistical analysis (for example, $p<.05$), which has permeated American ornithologists' discourse in recent years. Thus, the sharp rises in technical and polynomial technical terms in ornithology after the mid twentieth century, as well as the rise in statistical argumentation, clearly indicate the changes taking place in ornithological discourse at that time and are reflected in the eventual decrease in article length.

Regarding the volume of scientific publication, the three American astronomical journals studied by Abt (1981) registered ten-fold increases in the number of papers published between 1910 and 1980. A study of *The Auk* eventually confirms an increase in the number of published articles, but only by the 1970 volume, and then as a result of the journal allocating more space to article publication. Significantly, though, from 1884 to 1960, the number of articles published in *The Auk* continually *declined*, down by almost half in the 1960 volume, despite at least small increases in the space allocated to articles in all but two of the volumes studied. The 1980 and 1990 volumes recorded almost two-fold increases in the number of articles, less than the volume increase in astronomers' publications, but nevertheless significant. As stated above, in seeking parallels between changes in discourse and professionalization of a science, the timing here is of importance, because the increased space given to articles in *The Auk* between 1960 and 1970 again corresponds to the period identified by this study as the beginning of American ornithology's professionalization. That the volume of increase did not parallel astronomy's does not lessen the value of ornithology as a science. It merely indicates that astronomy was a more firmly established science. Abt suggested that the increase in astronomy articles paralleled the 28% increase in theoretical topics being discussed by astronomers. However, as chapter 5's overview of topics and methodology has shown, theoretical topics do not show marked increases in later *Auk* volumes and thus cannot be responsible for the increased volume of ornithologists' publication.

Use of Citations

Both physicists' and ornithologists' use of citations increased dramatically over their respective study periods. However, Bazerman (1988) found that physicists' references were eventually dispersed throughout articles and were discussed more thoroughly. The function played by citations in American ornithologists' writing is more problematic because, as a general rule, citations continued to be dispersed throughout articles. Ornithologists used citations almost uniformly over the years, first in early ornithology for citing specific locations of bird discoveries or physical descriptions of birds, and, later, for referencing specific empirical data. In either case, there was little discussion. Early *Auk* articles up to 1920, when the natural history phase prevailed, used a very loose form of reference, which might be termed "pseudocitations," an anonymous referencing based either on hearsay ("we were told by the Indians and others. . ." [Townsend & Bent, 1910, p. 7]) or on some named, presumably local, personage ("Mr. R. H. Lawrence states. . ." [Kobbé, 1900, p. 11]). Articles sampled in the 1920s more uniformly associated authors with referenced data—there were still no Literature Cited sections. The 1930s sample showed an inclination to group more citations toward the beginning of the article, after the manner of a literature review. Prior to this time, citations were generally dispersed throughout articles. By the 1940s, however, citations were again fairly dispersed and remained so until the 1980s, when more citations were transferred to the beginning of articles and can more strictly be called untitled Literature Reviews. But even in most recent *Auk* publications, authors continued to cite other scientists throughout the text. The emphasis upon observation throughout American ornithological research history, for which a simple citation suffices, along with limited theoretical argumentation, may account for the absence of lengthy paragraphs heavily embedded with citations. The inclination to disperse citations throughout American ornithological discourse suggests the reliance of a young emerging discourse on authoritative sources as evidence. The resurgence of markers for reliability of evidence in 1950 during the period of discourse interplay supports this suggestion by documenting further attempts to validate authority in preparation for American ornithology's approaching professionalization.

Sentence Length and Complexity

Unlike physicists' discourse, the sentence length and complexity of American ornithologists' discourse have reversed themselves during the field's development as a professional science. The pattern previously described for ornithology of a rise in article lengths until the mid twentieth century applies also for changes in average sentence lengths used by ornithologists. The pattern, though not as pronounced, is nevertheless clear. Like physicists' discourse, average sentence lengths remained relatively stable, but only for half a century until 1940, after which lengths measured a slow but steady decrease. The reverse trend was true for ornithologists' use of simple sentences, which declined after a peak in 1890 until 1940, after which percentages were above average except for 1970, but quickly rose in 1980 and 1990 to their highest averages. Here again, the period that this study has termed "discourse interplay" (1940–1970) appears to be the critical period for these two features of discourse, because usages reverse themselves during this time frame. Decrease in sentence length and increase in sentence simplicity are also predicted by the discourse models presented in chapter 3, which show longer sentence lengths and fewer simple sentences for natural history discourse, prominent in early American ornithology. Article lengths from the 44-article *Auk* database also parallel this ultimate decrease in sentence length as articles increased from 3,000 words on average to 9,000 by 1950, but steadily dropped to an average of 6,000 words by 1990.

Increased professionalization appears to be the most logical reason for declines in sentence length and complexity. Increasingly formalized technical terminology, theoretical underpinnings, methodological approaches, and recognized *topoi* enable the professional writer to say more in fewer words with fewer explanations. Halliday (1993a), for instance, has commented on the relative simplicity of sentence structures in scientific discourse (p. 67). Before professionalization, when American ornithologists had to communicate with both the amateur and "technical" enthusiast without the convenience of referring to accepted paradigms or terminology, lengthy explanations were needed to clarify positions and theoretical stances. In addition, an essential component of natural history discourse is narration and description, which of their nature involve more lengthy passages. Finally, the many anthropomorphic and sometimes emotional descriptions prompting early amateur-ornithologists to lengthy literary flights of style as excerpted in chapter 5, make clear the reason for lengthy, complex sentences in early American ornithology.

In the realm of sentence complexity, physicists used fewer relative clauses but more noun clauses and temporal/causal subordinate clauses, explained again by Bazerman (1988) as reflecting increased intellectual complexity. Here too, ornithologists' discourse was very different. Almost across the board, clausal usage changed very little over *The Auk*'s 100-year history, as shown in the subsample of *Auk* articles. Relative noun clauses and subordinate time, causal, and conditional clauses were relatively stable, whereas relative adjective and subordinate place clauses registered only small decreases in usage over the period, by two to three occurrences on average.

Thus, it is shown again that increased theoretical complexity does not necessarily mean increased discourse complexity. Though American ornithology did become more theoretically and intellectually complex during its 100-year history, the reason for this increased sophistication was the rise of statistical inference. In fact, statistical methodology has been an important player in the professionalization of American ornithology, because by its exclusionary nature it has increasingly restricted participation by amateurs who do not have the professional training either to implement the methodology, conduct the statistical analysis, or even understand the literature itself. Statistics has likewise played an important role in the changing discourse of American ornithology. As a mathematically symbolic means of argumentation, statistics can be used with minimal English. Such abbreviated expression perhaps accounts for the lack of significant changes in usage of noun and subordinate clauses by ornithologists. Much information can be imparted to a professional audience with a string of simple sentences loaded with p-values. The sample page from the 1990 *Auk* shown in Figure 5.10, for instance, demonstrates that Kerpez and Smith's many p-values permit simple sentence constructions. Thus, a science like American ornithology, which professionalizes in the midst of a statistical revolution, will not need to resort to changes in clausal structures in order to preserve accuracy in its discourse amidst more theoretically and intellectually complex knowledge.

However, one aspect of American ornithologists' style that has become more complex is the developing complexity of noun phrases, shown by the sharp rise in the use of words interrupting subject and verb, a feature seldom referenced in other studies of scientific discourse. Alan Gross (1990) included some examples of this construction in his discussion of scientists' preference for passive voice, which privileges noun phrases; but his emphasis was on the complexity resulting from a single scientific term expanded by descriptive modifiers as, for instance, "formalin-fixed Staphylococcus aureus cells" (pp. 70–74). Ornithologists, on the other hand, show preference for adding modifiers to any subject, not just technical ones, for instance, "The third, which Wolf calls the Subandine, but for which a continental-wide view of the subject suggests the name of Temperate Zone as more appropriate, is . . ." (Chapman, 1917, pp. 85–87). Although this type of expansion may be a unique feature of American ornithologists' writing, it more likely shows a greater need for definition, a major explanatory device in theoretical prose.

Word Choice

Both physicists' and ornithologists' discourse confirm that increases in the use of technical terms and multi-word noun phrases are key indices of mature scientific discourse. Halliday and Martin (1993) have clearly shown the origin, development, and impact of these two linguistic features on scientific discourse. Though technical terms are not difficult in themselves, they imply complex relationships and larger frameworks of thought (Halliday, 1993b, p. 71). They are distillers of thought since they "both compact and change the nature of the more commonsense meanings by which they are defined" (Halliday & Martin, p. 30). They are, in effect, a "kind of shorthand . . . enabl[ing] a more concise communication among specialists" (Wignell, Martin, & Eggins, 1993, p. 163). Similarly, nominalizations are a grammatical shorthand. They enable scientists to condense taxonomies (Halliday, 1993a, p. 64). They turn happenings into things (Halliday, 1993b, p. 82). They are "grammatical metaphors," which enable scientists to turn an entire set of experimental procedures into a single clause (pp. 79–81). In effect, their conciseness decreases negotiability (Halliday & Martin, pp. 39–41), thereby increasing scientific authority. Myers (1990) has noted the many subtle changes in meaning created by two biologists as they changed the diction of their proposals (pp. 51–52). Killingsworth and Palmer (1992) found ecologists' use of technical terms to be even more complex. In reality, anthropomorphic-sounding terms were actually loaded terms having specific scientific meanings for ecologists (p. 114). In this instance, terminology that appears to be non-technical is deceptively complex and scientifically context-bound.

In ornithology's case, sharp increases in technical terms and multi-word noun phrases beginning in 1950 show these terms to be key markers for identifying the professionalization of a science. Bazerman (1988) also found increases in substantive active verbs over reportorial ones. American ornithologists likewise increased their use of substantive verbs, reflected in the decrease in use of *be* verbs, despite the corresponding increase in passive voice. Writers, however, did not abandon reportorial verbs during the process of professionalization, a fact largely due to ornithologists' use of observation as an essential component of their methodology. Fueling the preservation of reportorial discourse in American ornithology has been the ever-present bird list, a process by which the observer's name and reputation have been inextricably attached to the observation itself. Thus, the truth of an observation is not determined just by the answer to "What was observed?" but also by "Who observed it?"

Graphics

A survey of English Renaissance texts shows the increasing use and effectiveness of visual techniques and page design (Tebeaux, 1991a; Tebeaux, 1991b; Tebeaux, 1997, pp. 35–89, 175–236). Ultimately, changes in Renaissance graphics from more primitive woodcuts to copper engraved plates and sophisticated line drawings revolutionized textual presentation (Tebeaux & Killingsworth, 1992, pp. 15–16; see also Tebeaux, 1997, pp. 175–236). Consequently, documenting change in graphical presentation of data over time is a simple but effective litmus test for gaining insight into several key questions: To what extent has a science professionalized; what are the ongoing changes in the methods and technology of data collection; and, more importantly, what constitutes scientific "knowledge" itself?

Bazerman (1988) found physicists choosing more abstract, technically sophisticated graphics over time. The same can be said for American ornithological discourse, which by the mid twentieth century was displaying its data in many more varied formats, such as multiple line and bar graphs, scatterplots, histograms, sonograms, and cladograms. Occurrences of all these graphics increased significantly after 1960, which this study has identified as the chronological turning point in the professionalization of American ornithology. The change from low-density graphics like lists to high-density graphs reflected the increasing amount of knowledge being transmitted by the discourse as ornithological science advanced. In effect, the use of higher-density graphics increased the number of trained readers, that is, researchers seeking detailed data to use in conjunction with their own findings (Killingsworth & Gilbertson, 1992, pp. 60–61). The essential role that graphics have played in the evolution of professional ornithological discourse is especially apparent in the presentation of statistical information, which has become the cornerstone of contemporary scientific rhetoric. The rise of statistics has been accompanied by sophisticated graphical presentation, which has become a powerful force in the persuasion of audience, as well as in its exclusion. Changes in the use of graphical presentation from photographs and line drawings to tables, line graphs, sonograms, and other graphical devices reflect corresponding changes in the rhetoric of ornithology. As Gross (1990) pointed out, tables and figures are important in scientific writing because they suggest relationships among the physical objects upon which science focuses (p. 74).

However, by studying the types of graphics appearing in *The Auk*, we discover a more complicated scenario. Bazerman (1988) found graphs replacing tables in physicists' discourse; but, in American ornithology, tables not only continue to be used but also increase more rapidly than graphs throughout the field's professionalization. Another interesting feature is that by 1990 bar graphs have superseded all other forms of graphs. Some technical writing texts, on the other hand, suggest that line graphs are more appropriate for the expert (or professional) audience, whereas bar graphs can be directed to any audience because of their ease of interpretation. Professional ornithology, at least, has not followed this prescription. The persistence of tables and bar graphs should also be interpreted in light of the continued popularity of maps and drawings in ornithology, as well as the unusually high use of photographs in 1970. Given that maps, drawings, and photographs are popular graphics in natural history discourse, their continuation late in ornithological discourse history, along with bar graphs and tables, may indicate that a completely professional American ornithological discourse is not yet a reality.

Organization and Formatting of Articles

For physicists' discourse, methods sections were prominent early in the century, with discussion, conclusion, and theory sections eventually dominating. Bazerman (1988) considered the increasing use of discussion and conclusion sections after 1930 to indicate more issue-oriented discourse. By contrast, early ornithology's methodology of field observation and collection was a well established procedure, the simplicity of which permitted the participation of the amateur field naturalist and thus required little explanation. Instead, its discourse was very much data-driven, organized around field and laboratory findings. Since ornithology's emphasis upon observation has never really subsided, its discourse has heavily emphasized results and discussion sections throughout its development. Still, as both observational and laboratory methodology became more sophisticated by the mid twentieth century, methodology sections were correspondingly more elaborate. It may be surmised that these sections evolved so differently from physicists' discourse as described by Bazerman because physics was already an established profession by the time the *Physical Review* had begun. On the other hand, American ornithology was actually developing into a science, which needed to establish a set of methodological procedures acceptable to the community. Thus, ornithologists' discourse, as it was being published in *The Auk*, helped to establish these methodological paradigms. As shown in chapter 5, American ornithology also became more issue-oriented after

1930, though natural history topics still maintained a foothold for two more decades. In contrast, theory sections have been almost nonexistent. Even when theoretical discussions or implications have been presented, they are usually sandwiched within discussions of factual data and presented in a few sentences. Thus, the paucity of elaborate theory sections signals American ornithology's youth as a professional science.

A most revealing distinction between ornithology and physics is the appearance of standard scientific formats in the respective field. In physics, section headings were established features after 1950, though Acknowledgments sections were not common until 1960. Again, in ornithology, the reverse is true. Indicative of American ornithology's youth and the influence of amateur narrative writing, section headings were very much delayed. There was, to be sure, frequent use of informative headings that described paragraph content. Again, because of ornithology's emphasis on observation, these author-invented headings focused on an observed behavior, object, or physical feature. The importance here of the appearance of standard section headings is that their presence is indicative of a field's organization, or professionalization. Their presence symbolizes scientists' communal sense of how their discourse should be structured, what they consider to be its essential components, and what is needed to lend credibility to a scientist's description of research. By 1940 and 1950, only a scattering of standard section headings appeared in *The Auk*, and then generally only one section heading used by a given ornithologist. Though all writers were using some section headings in 1970, the uniform use of a format that could be classified as "American ornithology's format" did not appear until October 1980. It may be inferred then that established formats appear late in a field's professionalization.

But there may be another interpretation of this delay. Gross (1990) traced the establishment of a scientific report format to Francis Bacon, who emphasized the inductive process mirrored in the format itself. According to Gross, "the experimental report has remained stubbornly ritualized, essentially unchanging" (p. 91). In addition, many technical writing textbooks likewise emphasize the presence of similarly inflexible established formats. This study of ornithologists' discourse, however, reveals a much more fluid structure than is often described. The establishment of a standard format for scientific discourse *is* a marker of professionalization; however, in American ornithology's case, the standardization arrived relatively late, not until 1980. If scientific discourse has an established format, the question to ask is why didn't ornithology quickly accept it as it *first began* the process of professionalization in 1960. The answer perhaps is that the uniform standard format for scientific discourse described by textbook authors may not actually exist in a fully prescriptive form.

A particularly revealing feature in the study of ornithological formatting is that, though Acknowledgments sections were not common in physics until after section headings were established, the Acknowledgments section in American ornithology was the *first* section to become established. It began appearing with frequency in 1960 and by 1970 was generally used along with Summary and Literature Cited sections. Given the reality of social forces at work in science, as shown by Ziman (1968), Kuhn (1970), Merton (1973), Bazerman (1988), Gross (1990), Myers (1990), and others, the recognition of one's peers is not only important in the maintenance of a scientific field, but most important in its professionalization. That ornithologists recognized this element early on is perhaps a crucial component, along with the rise of statistical inference, in the rapid professionalization of American ornithology after the mid twentieth century.

Co-Authorship

A communal awareness is also evident in the rapid rise of co-authorship among ornithologists. Though co-authorship in American ornithology has always been a viable alternative, co-authored articles became much more frequent during the period of discourse interplay between 1940 and 1970. By the last two decades, co-authorship had risen to over 60%; and three-, four-, five-, and even 10-author articles constituted 25% of *Auk* articles. By comparison, the Physics Survey Committee (1973) reported a modest rise in co-authorship among physicists from an average of almost two authors in 1936 to almost three in 1971. Multi-author articles of up to five contributors were also common among astronomers by 1980. Thus, though the volume of ornithologists' published articles did not approach that of astronomers' by 1980, the amount of co-authorship did. It appears, then, that, like the use of Acknowledgments sections, co-authorship was quickly accepted as American ornithology professionalized.

Responsible at least in part for this more social orientation is ornithology's early emphasis upon field observation. Many sciences, like physics, originated in the laboratory, where individual inspiration and performance were primary. But early American ornithology, primarily a field science, relied more heavily on a network of observers to communicate locations of study areas, as well as physical descriptions of species and their behavior. Ornithologists also shared access to museums, which housed rapidly accumulating collections of study skins. Because early ornithological science was shared with amateurs, whose interest in birds was often personal and nonscientific, it was essential to know "who" reported new discoveries since expertise varied greatly even within the "technical" ornithological community. The early science was also shared with conservationists and bird-lovers, whose interest in birds was often emotional and anthropomorphic. This group was likewise interested in ornithologists' field research, not for professional purposes but for the chance to observe new species being discovered and described yearly. Thus, from its inception American ornithology had an immediate popular audience with whom it could communicate its research on a nonscientific level. This association remains even today: Universities like Cornell solicit amateur assistance; and ornithologists continue to author bird-finding guides for the general public, probably the most obvious kind of popularization of ornithology.

In discussing the *topoi* of scientific reasonableness, Prelli (1989) described a scientist's mind set in this regard:

> Generally, scientific audiences are likely to judge a discourse unreasonable if they think the rhetor lacks the requisite qualities of thought and conduct needed to adduce scientifically reasonable aims and claims. . . . For instance, any member of an academic community has encountered charges that some members of the discipline are "popularizers" and not "serious scholars." (pp. 126–127)

Ecologists' discourse supports Prelli's assertion, since scientific ecologists distanced themselves from Leopold, Carson, and Commoner because they tried to extend scientific pedagogy to the general public.

American ornithology is perhaps an exception to this generality because of both the persistence of amateur involvement and popular interest in ornithology's various research agendas. For instance, Cornell's Arthur Allen still became an eminent ornithologist despite his popularizing of birds. Nor was Frank Chapman's inclusion into the ranks of the professional ornithologist hampered by his own popularizing of bird study, either through his museum displays for the general public; through his *Handbook*, which he acknowledged was written for audiences other than the professional; through his intimate association with the National Audubon Society; or through his invention of the Christmas bird count, perhaps the one event that has popularized the study of American ornithology more than any other activity. During its nearly 100-year existence, the count has been the most prominent link between amateur and professional, most evident in the decision by the prestigious Cornell Laboratory of Ornithology to begin computerizing long-standing Christmas bird count data for use in scientific research. In a related effort, Cornell's Lab solicits assistance annually from amateur field observers for such research activities as nest data collection, winter feeder-bird surveys, winter habitat studies, and seed preference projects. Such activities show an interest in forging a cooperative effort to unite both amateur and professional under one scientific roof. On the negative end, these "volunteer" projects often require monetary contributions by the amateurs whose help is solicited. As a result, such financial agendas may be regarded as a sign either of the communal forces that have united ornithology throughout its history or of the commercial state into which ornithology as an amateur-professional venture has fallen.

Scientific Argumentation

In his study of argumentation by physicists involved in spectroscopy, Bazerman (1988) found that physicists' argumentation during the century evolved from explanation of methodology to theory- and problem-oriented articles, the theoretical side restricted after 1930 to considerations within quantum mechanics. In the 1950s and 1960s, theory-testing gave way to more tentative model-building.

American ornithology's argumentation developed somewhat differently, particularly with respect to methodology. In fact, its development actually closely paralleled *stasis* procedures (see Fahnestock, 1993, pp. 31–34; see also Prelli, 1989; Gross, 1990). Unlike spectroscopy, early American ornithology's methodology was well established: use of the gun to discover and collect bird species. Ornithological research therefore dealt with new and geographically variant forms and subspecies of birds; that is, argumentation focused on stasis questions dealing with existence (Is it a new species or not?), meaning (What does it look like?), and quality (Where in the taxonomy does it belong?). Throughout its history, ornithology has been uniformly problem-oriented, proposing identifications of birds and describing their habits, behavior, and life cycles. But, unlike spectroscopy discourse in physics, methodological argumentation in American ornithology took on greater importance as the science matured. Thus, the last *stasis* procedure (What action to take?) gained in importance.

Early in the century, new methodology, such as bird banding and basic statistical procedures, needed aficionados. By 1940, field studies were becoming more manipulative, requiring detailed methodological explanation. By mid century, when American ornithology began to professionalize, new sophisticated equipment and statistical procedures with their accompanying experimental designs required explanation, followed closely by theoretically complex, controversial DNA methodology. Thus, throughout the history of American ornithological discourse, methodology sections required more, rather than less, space.

Paralleling spectroscopy, however, model-building did increase in American ornithology, though somewhat later, appearing more frequently after the mid twentieth century. Ornithology, however, was not given over to this type of argumentation, because its problem-oriented natural history discourse still persisted. Theoretical argumentation did not become prominent in ornithological discourse until the last two decades with the predominance of statistical procedures. Earlier, theoretical argumentation was restricted primarily to bird migration, at least in the discourse analyzed in this study. So, unlike traditional schemas describing the development of scientific knowledge from methodology to theory- and problem-solving, argumentation in American ornithological discourse was not a direct progression from simple to more complex forms. Problem-solving argumentation was uniformly present; methodological argumentation gradually increased, whereas model-building and theory-oriented argumentation have registered recent births.

It should be noted here also that century-old taxonomic issues continued to appear in ornithologists' research agendas. Prelli (1989) remarked that "Problems that occur when scientists try to fit individual cases or sets of data into existing taxonomies are among the most representative cases of definitional *stasis*. . . . The problem does not lie in the taxonomic principles and classification schemas themselves" (p. 149). But this is not true for American ornithology; their problems and discourse have often to do with altering taxonomies. That process is part of doing ornithology.

Gary Gutting (1980) has observed that a sign of a mature science is one with a consensus that "eliminates the need for further discussion of foundational questions about the subject-matter and methodology of the disciplines and enables the discipline to devote most of its energy to puzzle-solving" (p. 13). The decrease in methodological argumentation in American ornithologists' discourse in 1980 and 1990 perhaps reflects Gutting's observation, another verification of American ornithology's professionalization after 1970.

Marcello Pera (1994) provided an alternative view of such changes in argumentation. An application of his theory of scientific discourse to ornithology would posit that ornithologists' straightforward methodology of gun and specimen gave way to more complex methodologies such as statistical and biochemical analysis, the complexities of which require vaguer rules and more context-specific methodologies and the explanations of which require rhetoric. Thus, article lengths and methodology sections increased during much of *Auk* history. Now professionalized, American ornithological science will move from explanatory to interpretative theory; and ornithologists' discourse should again revert to more complex, lengthy argumentation.

CONCLUSION

It is not possible to compress an evolving scientific discourse into neat Latourean black boxes that identify discourse changes as a science professionalizes. Instead, the influence of novice voices—that is, amateur scientists—who add their own variation to a developing discourse must also be taken into account in order to understand fully the processes at work in the entire cycle of a developing scientific discourse. Experimental parapsychology, for instance, suffered professionally because of its amateur practitioners (Allison, 1979). The role of the amateur was eventually institutionalized within British astronomy (Lankford, 1981, p. 11). Amateurs were needed in British geology until after World War II, though only for financial support (O'Connor & Meadows, 1976). But as Taylor (1996) has pointed out, the literature on the professionalization of science does not describe any explicit attempt to include or exclude the amateur (pp. 95–96). Nor has this study found any conscious exclusionary pressures from within American ornithology. Too often, discussions of the professionalization of science overemphasize the incompatibility between amateur and professional. Science is seen as an exclusionary body, consciously "throwing out" the amateur scientist. But some fields, like ornithology, have benefited from amateur and professional associations. Rather, it is through technological sophistication, not conscious effort or desire, that the split between amateur and professional has been made.

To fully understand the professionalization of a science, its discourse must be studied from the field's inception, a particularly challenging process for sciences like botany, physics, and chemistry, which have such protracted periods of discourse and professional histories. The study of spectroscopy articles by Bazerman (1988) in the *Physical Review*, for instance, was a case study of scientific discourse in process because physics was already an established profession. By 1893, the year *Physical Review* began, the science of physics had already proposed scientific theories concerning mechanics, optics, light, electricity, magnetism, and thermodynamics, to name a few. Ornithology had no such history of contributions to scientific knowledge in 1883 when the 21 interested "ornithologists" gathered to form an organization. Perhaps, ornithology's most significant impact upon science up to that time was the contribution made to evolutionary theory by Darwin's study of the Galapagos Island finches. As further evidence of the obscurity of the meaning of "ornithologist," Chapman (1935) had left his well-paying bank job in 1886 to become one, but he admitted in his autobiography that he didn't even know what it meant. All he had envisioned was that his methodology would be to observe and his data would be a bird list. He had no idea what he would do with the information, once he gathered it.

This study of American ornithological discourse, then, has delineated the processes at work in the emergence of the professional discourse of a science from origin to maturity. Other young emerging sciences should take note of the necessary intermingling of the interested amateur and the knowledgeable expert, both of whom can contribute usefully to the developing discourse. Incorporating research agendas, topics, and theories from other already established sciences has the advantage of adding *topoi* and paradigms, which can hasten the maturation of an emerging scientific discourse. At least in American ornithology's case, when contemporary discourse matures, it does so rapidly, not with complex discourse but rather from discovery, development, and usage of the tools that make the discipline a unique science. These tools of the trade transfer into the discourse through technical terminology, invented expressions through nominalizations, expansions of subjects, and sophisticated visual displays of data. Wordiness gives way to conciseness, explanation to inference, and the individual writer to co-authorship.

The unique, and sometimes contradictory, development of American ornithological discourse reflects both the persistent influence of the amateur in American ornithology and the slow evolution toward professionalization, which this study of *The Auk* demonstrates—changes due to an evolving professional audience, to more defined paradigms, to a rise in statistical inference, and to more sophisticated instrumentation and methodology. Thus, a study of American ornithological discourse sheds further light into the development of scientific discourse because it shows the progression of a science, American ornithology, *from its inception*, whereas studies like those by Bazerman (1988) and Abt (1981) are entering physics and astronomy discourse *in media res*. This study of American ornithologists' discourse shows also that the influence of the novice voice of the amateur scientist in a developing discourse must also be taken into account in order to understand fully the processes at work in a developing scientific discourse. The study provides an opportunity to see the changes created as a science develops a sense of its own uniqueness as a profession and begins to cement the bonds that establish it as a unique, paradigm-bounded community of scientists, whose discourse reflects their sense of professionalism. The study also suggests other avenues for research:

- How might we describe the development of other fields, like dialectology, the origins of which, like ornithology's, have depended upon observation and collection?
- How different are the discourse models operating in other scientific fields like ecology and environmental science, the direct impacts of which upon society have necessitated more rhetorical expertise than that needed by ornithology?
- What do the discourse schemas of other scientific fields look like?
- What continuums operate within other scientific fields?
- What would historical studies of other twentieth-century sciences besides ornithology and physics tell us?
- Given the technological and statistical sophistication of the Information Age, could a newly emerging science simply bypass its amateur stage?

This study of *The Auk* and American ornithology then contributes additional assertions about the processes at work in a developing scientific discourse through a study of an evolving discourse coupled with a jointly evolving professional persona. The study ultimately shows that much more research is needed before we can place the paradigms of scientific discourse into our own Latourean black boxes.

Appendix A

SAMPLE PAGES CODED FOR LINGUISTIC ANALYSIS OF THE AUK

436 HARVEY I. FISHER [Auk, Vol. 92

TABLE 2

MORTALITY OF YOUNG, BREEDING LAYSAN ALBATROSSES IN THE PLOT

Sex	Original number	Percent[1] during breeding year 1	2	3	4	5	6	7	8	9	Mean
Males	531	2.1	4.0	6.8	5.5	5.6	5.4	1.3	2.7	2.7	4.0
		2.1	4.0	7.2	6.3	6.9	7.2	1.9	3.8	4.0	4.8
Females	429	2.6	3.7	4.0	3.7	6.0	4.9	2.0	3.9	2.8	3.7
		2.6	3.8	4.2	4.2	7.0	6.1	2.8	5.4	4.1	4.5

[1] Percent is calculated in two ways: the first line is percent of original sample, the second is percent of birds alive at beginning of risk period.

METHODS

The methods and techniques ~~used to obtain the data on Midway Atoll~~ may be found in Fisher and Fisher (1969) and Fisher (1971). Of the two basic samples, one was of chicks banded in the study plot between 1956 and 1963, inclusive, and found breeding there between 1961 and 1972. Sexing ~~at the time of recapture~~ showed 531 males and 429 females. The second sample contained 1068 birds of the 1960- to 1963-year classes, that ~~first~~ bred between 1966 and 1972. Available assistance did not permit sexing these birds.

The interpretation and other use ~~of the data on mortality~~ require a word of caution. The Midway Atoll colony ~~of Laysan Albatrosses~~ has grown greatly in numbers since the turn of the century, and breeding numbers have approximately doubled in the last 15 years (Fisher MS). Consequently all ~~the observed mortality~~ rates may be atypically low for the species as a whole, but this possibility should not affect the validity of comparisons between rates observed for various cohorts within the same span of years—the use made of them in this study.

RESULTS

Mean annual mortality (~~percent of original sample~~) was 4.0% (1.3 to 6.8%) for young males in their first nine breeding seasons and 3.7% (2.0 to 6.0%) for females (Table 2). In an unpublished study I have demonstrated that the mean annual rates ~~for birds of all breeding ages in two colonies~~ were 5.3 and 6.3%, varying annually between 1.9 and 8.5% and showing no significant sexual variation during the 13-year period. If mortality is expressed as percent of birds exposed to risk in each year (Table 2), the mean annual rate ~~for young breeding males~~ was 4.8% (1.9 to 7.2%) and 4.5% for females (2.6 to 7.0%).

Both expressions ~~of mortality~~ revealed that breeding birds ~~of both sexes~~ were much more vulnerable from the third to the sixth breeding years, inclusive. Annual mortality ~~in breeding years 1 and 2 and 7, 8, and 9~~ averaged approximately half that during years 3 to 6.

As males begin breeding essentially one season earlier than females, the survival rates ~~in Table 3~~ must be offset one year for comparison.

The rates ~~of survival of 9-year-old males and 10-year-old females~~ show

FIGURE A.1. Sample text page coded for linguistic analysis (text: Fisher, 1975, p. 436).

AUTHOR	WFN					ID 1975.1		
AVE. SENTENCE LENGTH (|) 19.8 1051/53		TOTALS	Intro	Methods	Results	Discussion	Summary	
METADISCOURSE (M) (writing about the reading or writing)								
(we claim, to summarize...)	rf. to own thinking 1							1
(first, more imptly...)	rf. to structure of writing 2							2
(note, to understand..)	reader's act of reading 3							3
DICTION								
	NOMINALIZATION (N)	14	HHT ///	/	/	//	//	(N)
	TECHNICAL TERM (T) 1	2		//				(T)
	polynomial (PT) 2	7	/	/		HHT		(PT)
	PERSONAL PRONOUN (PP)							(PP)
	DEFINITION (D)							(D)
SENTENCE COMPLEXITY								
(it, one, there)	subject as place holder (P)	2			/	/		(P)
	Passive voice (PV)	9	///	////	/	/		(PV)
	to be (checked) 1	14	/	////		///	HHT /	1
(she stated, etc.)	reporting verbs (checked) 2	1				/		2
	substantive active verbs (checked) 3	69	HHT HHT //	HHT ////	HHT / HHT HHT HHT	HHT HHT /	HHT HHT HHT	3
	no. simple sentences / total no. sentences	26/53						#
Complex Clauses (boxed)								
	rel-n 1	~~5~~ 8	/		//	//		1
ao-11T +11b } #1 (1885.1)	rel-a 2	12	////	/	///	///	/	2
	s-time 3	2			/		/	3
ao-146	s-causal 4	1					/	4
ao	s-place 5	1				/		5
(if, though, unless, etc.)	s-condx 6	2	/	/				6
Subj-verb Interruptions (strike-through)								
	no. / total no. possible 1	25/83	HHT /	///	//	HHT ///	HHT /	1
	no. interrupting words 2	175	4,10,4,11,10,3	6,4,5	17,3	4,5,19,10,9,5	11,2,5,4,4,12	2
EVIDENTIALS (linguistic markers for making factual claims) (EV circled)								
	RELIABILITY OF EVIDENCE (ER) 1	22	//	////	////	HHT	HHT //	(ER)
(basically, by definition, certainly, essentially, exactly, generally, in some sense, invariably, literally, may, maybe, might, normally, particularly, perhaps, possibly, primarily, probably, specifically, surely, undoubtedly, virtually)								
(I guess, suppose, think)	BELIEF (EB) 2							(EB)
(evidently, must, obvious, seem)	INDUCTION (EI) 3							(EI)
(apparently, I'm told, people say, supposed to, it seems)	HEARSAY (EY) 4							(EY)
	CITATION (C)	11	HHT	/		HHT		(C)
(can, could, presumably, should, would)	DEDUCTION (ED) 5	1					/	(ED)
(about, kind of, sort of)	HEDGING (EH) 6	7	////				///	(EH)
(feel, hear, see, feel-look-sound like)	SENSORY EV (ES) 7							(ES)
(actually, at least, but, even, however, in fact, nevertheless, oddly enough, of course, only)	EXPECTATIONS (X) 8	10	/		////	//	///	(X)

FIGURE A.2. Sample of a completed coding form.

Appendix B

ARTICLES USED IN CHAPTER FIVE FOR THE LINGUISTIC ANALYSIS OF THE AUK

Allen, J. A. (1890). To what extent is it profitable to recognize geographical forms among North American birds? *Auk, 7,* 1–9.

Beer, C. G. (1965). Clutch size and incubation behavior in Black-billed Gulls (*Larus bulleri*). *Auk, 82,* 1–18.

Brewster, W., & Chapman, F. M. (1895). Notes on birds observed in Trinidad. *Auk, 12,* 201–211.

Brooks, A. (1920). Notes on some American ducks. *Auk, 37,* 353–367.

Broun, M. (1935). The hawk migration during the fall of 1934, along the Kittatinny Ridge in Pennsylvania. *Auk, 52,* 233–248.

Buckley, P. A., & Buckley, F. G. (1970). Color variation in the soft parts and down of Royal Tern chicks. *Auk, 87,* 1–13.

Christy, B. H. (1930). British birds at a glance. *Auk, 47,* 11–21.

Cobb, S. (1920). Midsummer birds in the Catskill Mountains. *Auk, 37,* 46–49.

Connell, C. E., Odum, E. P., & Kale, H. (1960). Fat-free weights of birds. *Auk, 77,* 1–9.

Cooke, W. W. (1905). Routes of bird migration. *Auk, 22,* 1–11.

Dunning, J. B., Jr., & Watts, B. D. (1990). Regional differences in habitat occupancy by Bachman's Sparrow. *Auk, 107,* 463–472.

Eifrig, C. W. G. (1905). Ornithological results of the Canadian "Neptune" expedition to Hudson Bay and northward. *Auk, 22,* 233–241.

Fisher, H. I. (1975). The relationship between deferred breeding and mortality in the Laysan Albatross. *Auk, 92,* 433–441.

Gill, F. B. (1980). Historical aspects of hybridization between Blue-winged and Golden-winged Warblers. *Auk, 97,* 1–18.

Greenewalt, C. H. (1955). The flight of the Black-capped Chickadee and the White-breasted Nuthatch. *Auk, 72,* 1–5.

Gromme, O. J. (1935). The goshawk (*Astur atricapillus atricapillus*) nesting in Wisconsin. *Auk, 52,* 15–20.

Groskin, H. (1945). Chimney Swifts roosting at Ardmore, Pennsylvania. *Auk, 62,* 360–371.

Henshaw, H. W. (1900). Occurrence of *Larus glaucescens* and other American birds in Hawaii. *Auk, 17,* 201–206.

Johnson, N. K., & Johnson, C. B. (1985). Speciation in sapsuckers (*Sphyrapicus*): II. Sympatry, hybridization, and mate preference in *S. ruber daggetti* and *S. nuchalis*. *Auk, 102,* 1–15.

Kennard, F. H. (1915). On the trail of the Ivory-Bill. *Auk, 32,* 1–14.

Kilham, L. (1960). Courtship and territorial behavior of Hairy Woodpeckers. *Auk, 77,* 259–270.

Kobbé, W. H. (1900). The Rufous Hummingbirds of Cape Disappointment. *Auk, 17,* 8–15.

Kroodsma, D. E., & Canady, R. A. (1985). Differences in repertoire size, singing behavior, and associated neuroanatomy among Marsh Wren populations have a genetic basis. *Auk, 102,* 439–446.

Miller, G. S., Jr. (1890). Cape Cod bird notes. *Auk, 7,* 226–229.

Mock, D. W., & Mock, K. C. (1980). Feeding behavior and ecology of the Goliath Heron. *Auk, 97,* 433–448.

Moreau, R. E., & Moreau, W. M. (1940). Incubation and fledging periods of African birds. *Auk, 57,* 313–325.

Paine, R. T., Wootton, J. T., & Boersma, P. D. (1990). Direct and indirect effects of Peregrine Falcon predation on seabird abundance. *Auk, 107,* 1–9.

Phillips, J. C. (1910). Notes on the autumn migration of the Canada Goose in eastern Massachusetts. *Auk, 27,* 263–271.

Reed, B. P. (1925). Growth development and reactions of young Great Horned Owls. *Auk, 42,* 14–31.

Richardson, F., & Fisher, H. I. (1950). Birds of Moku Manu and Manana Islands off Oahu, Hawaii. *Auk, 67,* 285–306.

Salomonsen, F. (1965). The geographical variation of the Fulmar (*Fulmarus glacialis*) and the zones of marine environment in the North Atlantic. *Auk, 82,* 327–355.

Scholander, S. I. (1955). Land birds over the western North Atlantic. *Auk, 72,* 225–239.

Schufeldt, R. W. (1915). The fossil remains of a species of Hesperornis found in Montana. *Auk, 32,* 290–294.

Schwartz, C. W., & Schwartz, E. R. (1950). The California Quail in Hawaii. *Auk, 67,* 1–38.

Scott, W. E. D. (1885). On the breeding habits of some Arizona birds. *Auk, 2,* 1–7.

Skutch, A. F. (1930). The habits and nesting activities of the Northern Tody Flycatcher in Panama. *Auk, 47,* 313–322.

Soper, J. D. (1940). Local distribution of eastern Canadian Arctic birds. *Auk, 57,* 13–21.

Sprunt, A., Jr. (1925). An avian city of the South Carolina coast. *Auk, 42,* 311–319.

Stephens, F. (1885). Notes of an ornithological trip in Arizona and Sonora. *Auk, 2,* 225–231.

Townsend, C. W., & Bent, A. C. (1910). Additional notes on the birds of Labrador. *Auk, 27,* 1–18.

Tyrrell, W. B. (1945). A study of the Northern Raven. *Auk, 62,* 1–7.

Verbeek, N. A. M. (1970). Breeding ecology of the Water Pipit. *Auk, 87,* 425–451.

Widmann, O. (1895). A winter robin roost in Missouri, and other ornithological notes. *Auk, 12,* 1–11.
Woolfenden, G. E. (1975). Florida Scrub Jay helpers at the nest. *Auk, 92,* 1–15.

Notes

Chapter 2

[1] There are few histories of ornithology. The most comprehensive study of European ornithology is Erwin Stresemann's *Ornithology: From Aristotle to the Present* (1951/1975), which includes a parallel chapter by Ernst Mayr on the rise of American ornithology. Paul Lawrence Farber (1982) discusses the development of European ornithology as a scientific discipline in *The Emergence of Ornithology as a Scientific Discipline: 1760–1850*. Readers interested in a thorough discussion of the development of early ornithological discourse in America are referred to *History of American Ornithology before Audubon* by Elsa Guerdrum Allen (1951), who meticulously investigated European and American manuscripts. Elliott Coues' historical overview in his *Key to North American Birds* (1903) shows how American ornithology of the late nineteenth century understood its own origins. More recently, Barbara Mearns' *Biographies for Birdwatchers* (1988) provides biographical sketches for the many ornithologists whose names are now parts of the common nomenclature for birds. Finally, two books trace the rise of popular ornithology in the United States: *Neighbors to the Birds: A History of Birdwatching in America* by Felton Gibbons and Deborah Strom (1988) and Joseph Kastner's *A World of Watchers* (1986), which also includes valuable chapter bibliographies.

[2] Wilson's work would never have been completed had it not been for George Ord, patron and co-founder of the Academy of Natural Sciences in Philadelphia. Ord oversaw the publication of volume 8 of Wilson's *Ornithology* after the death of the painter-ornithologist. Volume 9, containing a final 16 species, was actually the work of Ord himself. Though he used notes from Wilson's journals, he supplemented Wilson's drawings by his own fieldwork and by correspondence with other naturalists and museums. Ord was also important in the developing American ornithological discourse for his writing of the zoological sections of Gutherie's *Zoology*, "the first systematic work on

American zoology," which included several descriptions of new American birds (E.G. Allen, 1951, pp. 565–566).

[3] In addition to these painters, John White, a talented colonial bird artist, came to America with Walter Raleigh, for whom White produced three volumes of original water colors of American birds in 1585. However, White's work had little impact on American ornithological discourse since his art work was not published.

[4] Wignell, Martin, and Eggins (1993) demonstrate taxonomic classifications used by both amateur and professional ornithologists (pp. 137–143).

[5] Interestingly, though not surprisingly, early American ornithologists preferred to disavow their evolution from a former bird club, apparently suggestive of a less than scientific origin. Ornithologist C. F. Batchelder, one of the founding fathers of the American Ornithologists' Union, goes to great pains to argue against such a direct link. See *An Account of the Nuttall Ornithological Club: 1873 to 1919*, also reprinted by ornithologists Chapman and Palmer (1933) in the AOU's 50-year commemorative volume on American ornithology's progress. However, Batchelder is unable to explain adequately how the *Bulletin* happens to disappear just as *The Auk* emerges or why the cover of *The Auk* used as a subtitle "continuation of the Bulletin of the Nuttall Ornithological Club." Most telling is this sentence from the inside front cover of the issues published during the 1880s: "In general character [*The Auk*] differs little from the late 'Bulletin of the Nuttall Ornithological Club,' of which it forms virtually a second series" (*Auk*, 1894).

Chapter 3

[1] In a related analysis, Jean-Francois Lyotard (1984) distinguished another dichotomy: narrative and expository discourse (see Battalio, 1996). Narrative discourse, an essential component of natural history writing, is created by ordinary human beings recounting stories either that they have heard or in which they have themselves been participants. The speakers' right to speak these narratives comes both from community membership and from having something of interest to say. Expository, or scientific, discourse, however, is denotative, created by privileged speakers who have their own special claims to truth. Their listening community is also privileged in that its mere existence is determined by agreed upon methods and theories. Though both discourses function in community, narrative discourse creates truth according to community-sanctioned mythic patterns, whereas scientific discourse establishes its truth through method, which in turn requires learning and thus a teacher. This teaching function enables narrative likewise to appear within the scientific community, as part of its attempt to gain student-adherents to perpetuate the scientific community and expand upon scientific theory.

Thus, through narrative, scientific facts, under the guise of indisputable truths, are transmitted to would-be scientists, whose eventual mastery of scientific method may allow them to turn these formerly indisputable facts into new theories and methodologies. Because this scientific narrative is addressed to novices as a means of gaining adherents, a kind of initiation rite, Lyotard (1984) suggested that the scientific community in general devalues such simple narrative discourse, like that found in science textbooks and in television programs such as *Nova* (Killingsworth & Palmer, 1992, pp. 126–131).

Though Lyotard's description of the dichotomy between narrative and expository discourse in science may adequately describe the established sciences, the scientific writing of emerging sciences, like American ornithology, does not fit neatly into these categories. In contrast, narratives function differently in young sciences attempting to acquire adherents and make a niche for themselves in the realm of science.

In addition, numerous authors have pointed out the importance of narrative in the biological sciences. Ornithologist Ernst Mayr (1982, pp. 71–73) has discussed the importance of narratives in the argumentation of biologists. Frederic Holmes (1991) also recently argued that narrative is present in contemporary scientific writing, and Myers (1991) has shown the function of narrative in the scientific review article. Elsewhere, Myers (1990) argued that biological narrative may be merely implicit in the text (p. 81), whereas Halliday and Martin (1993) have shown how nominalization and grammatical metaphor in science enable stories to be told concisely.

[2] Myers (1990, pp. 72–73) summarized other methods for evaluating evidence claims in texts. See also Fahnestock (1993, pp. 21–30).

[3] In order to evaluate these linguistic features, the selections used to develop the three discourse models in chapters 3 and 4 and to represent American ornithologists' discourse in chapter 5 were analyzed as follows. The method for selection of passages for analysis (following Bazerman, 1988, p. 167) was the use of the first 200 words (ending with the first sentence break) from the first five major sections (Introduction, Methods, Results, Discussion, and Summary/Conclusions or corresponding section headings); if any section contained less than 200 words, the number of words needed to make 200 was divided equally among each of the remaining sections so as to obtain a sample of 1,000 words. If there were no headings, the first 1,000 words of the article were used.

The 32 features used for the linguistic analysis of each selection are listed in the tables accompanying each discourse model and are summarized in the sentence-level analysis section of chapter 3. Besides providing a comprehensive coverage of linguistic options from metadiscourse to sentence structure to word choice, these features parallel those used by Bazerman (1988) in his analysis of physicists' discourse (pp. 153–186). Thus, the ability to make direct comparisons between physicists' and ornithologists' discourse will enhance chapter 6's analysis of the role American ornithological discourse has played in the evolution of scientific discourse.

Data were collected by coding directly onto a copy of the text being analyzed and then transferring that data to a coding sheet, which was then tallied and the results averaged and transferred to tables representing each discourse model or to a spreadsheet program for creating descriptive graphics. The exact procedure was as follows. As each linguistic feature was encountered in the sample being analyzed, that feature was coded onto the text using the codes listed in parentheses on the coding sheet. (See Appendix A for a marked text page and sample coding sheet.)

At least three passes were made through each text, one pass each for marking word choices (sentence complexity and diction), metadiscourse, and evidentials, respectively. Nominalizations were marked if the writer could have rewritten the nominalized word in its verb form. Campbell & Lack (1985) were referenced if it was unclear whether a particular term was technical. The definition of metadiscourse was that provided by Williams (1994, pp. 79–81). Questions about grammar were resolved by referencing Jespersen (1964; 1909–1949) and Quirk (1985), in that order.

Then, the average sentence length and percentage of simple sentences were tallied. The data were next transferred to a coding sheet, which was then used to total the occurrences of each feature in the text under study. Finally, these totals were entered into tables for analyzing the discourse models and into a spreadsheet program to create descriptive graphs for analyzing American ornithologists' discourse in *The Auk*.

Because of a computer's limited capability to interpret such items as metadiscourse and evidentials, electronic coding was not used. Consequently, all texts were hand-coded and tallied by the author. Before coding the texts used in this study, a set of sample pages was coded and then recoded a week later in order to evaluate the coding form and procedure and to improve coding consistency. Materials were coded daily until completed.

[4] John Ziman (1981, pp. 123–137) provided additional insight into the answers to these questions in a chapter entitled "Why Be a Scientist?", the answers to which were based on the social context of science.

[5] This process, greatly oversimplified here, carries with it much controversy. See Sibley and Ahlquist (1986) for a more complete explanation of DNA-DNA hybridization and DeBenedictis (1988a, 1988b) for an overview of the controversy surrounding the methodology and its implications for the classification of birds.

Chapter 4

[1] Only the major articles in each volume were considered in this analysis. Though other discourse, such as notes, short communications, letters to the editor, obituaries, and American Ornithologist Union reports, are also potentially valuable sources in analyzing the development of American ornithological discourse, these were excluded from the survey.

[2] Mention should be made here of the methods used to classify the articles analyzed. (a) The amount of scientific sophistication described in the article determined its placement along the Continuum of Scientific Perspectives on Nature. That is, an article was placed farthest to the right as possible. For example, if an article described both field observation and collection of specimens, then it was placed in the more theoretical category "bringing system into lab." However, in general, I used the overall focus, intent, and expertise of the research described by the author, rather than one or two isolated events, to categorize an article. For instance, in "A Winter of Rare Birds at Ottawa, Ontario" (Eifrig, 1910), although the author references the stomach contents of Evening Grosbeaks and observes caged Pine Grosbeaks (i.e., "bringing system into lab"), these were only isolated references. The author's intent is to describe observations of species made during the winter of 1908–1909 in Ottawa. Thus, the article was categorized as "observe in field." The graphical displays discussed in this chapter will adjust for this

overlapping. (b) Articles which used counting methods in the field were classified in either of two ways. First, an article was classified as "observing in field" if counts made were casual counts (identified by expressions such as "about," "approximately," and the like), were of small numbers of birds (like groups of 3 or 4 which could be quantified merely by casual observation), or were referencing only numerically even numbers of sightings (like 1,000 blackbirds or 3,500 hawks). If specific attention was given to counting (like "347 starlings," "17 eggs," or "16 nest visits"), then the article was classified as "observing and measuring in field." (c) In the 1890, 1900, and 1910 volumes, which do not list addresses of writers and the individual articles of which do not indicate origin, an article was classified as "bringing system into lab" if it can be reasonably assumed that measurements described may not have been completed "in the field" (for instance, analysis of stomach contents or mm. measurements of testes). (d) "Manipulation with numbers" was not chosen if statistics were used only peripherally for support. For example, "Alternative Foods of a Diet Specialist, the Snail Kite" (Petersen, 1990) is an observation-based study in which the researcher is primarily observing the feeding behavior of kites and grackles. Though he references five p-values (on handling times [3], prey weight [1], and capture time [1]), these are of relatively minor importance considering the overall intent of the article. The majority of the Results section discusses his observations using percentages primarily, if he references numerical data at all. P-values are mentioned only in the last subsection and then relegated to the end of the section. His Discussion refers to inferences derived from these five p-values only in passing; and he does not use the words "significantly different," or other similar expression, by which he would be directly attaching statistical inference to his analysis. (e) The "other" category was used for 21 articles, such as those dealing with ornithological history, polemics, methodology, and descriptions of instruments, which do not fit into the Continuum. Whether or not this is the continuum most useful for ornithological science will be discussed in chapter 6; for instance, in early American ornithology, "observing and shooting in the field" may be instructive as an additional category. However, as will be seen, the Continuum describes the progress of American ornithology quite well and will clearly demonstrate the interplay between natural history and experimental science.

[3] Because the volumes surveyed were chosen by decade in time-ordered sequence in order to provide a historical perspective, the data to be analyzed cannot be considered a simple random sample. However, the relatively large number of articles should provide a useful indication of the progression of methodology and discourse in American ornithology.

[4] These 1980 articles are those beginning on pages 135, 148, 245, 521, 655, 704, 742, 760, 768, 795, 818, and 861; and in 1990, pages 61, 69, 75, 110, 161, 246, 473, 544, 649, 660, 669, 707, 718, 730, 737.

[5] These 1980 articles are those beginning on pages 19, 283, 351, 509, 559, 782, and 825; and in 1990, pages 18, 25, 86, 96, 133, 146, 334, 367, 376, 490, 500, 506, 580, 678, 689, 696, and 750.

[6] Myers (1990, pp. 141–192) has already identified the general features of both scientific discourse and popularizations of science. For his analysis he studied the writing of three biologists, who each published articles on similar subjects for

different audiences. A goal of the present study, however, is to look closer at specific observable and quantifiable discourse features as indices of discourse boundaries between amateur and professional writing. Also, whereas Myers (1990) used semiprofessional journals as examples of popularizations, this study uses strictly amateur discourse, since the intent is to study the professionalization of American ornithology.

[7] Though ornithologists' own professional writing can assist a researcher in determining worldviews and social perspectives useful for developing discourse models in ornithology, biographical and autobiographical accounts can also give additional perspectives. There are, however, not many biographical accounts of ornithologists and few autobiographies. Most notable are those by Chapman (1935), Alexander Skutch (1971), George M. Sutton (1980), Olin S. Pettingill (1992), and Frances Hamerstrom (1994). Concerned with projecting objectivity, scientists are not accustomed to expressing their personal thoughts and emotions on paper. It may be argued that comments made by Chapman in his autobiography or by his obituarists cannot be accepted at face value. That is, the reader cannot necessarily attach concepts of worldview, social perspectives, and identity of self to such accounts, under the assumption that the writer's complete honesty cannot be assumed. However, the fact that Chapman or his obituarist would make statements about himself or his perspectives or viewpoints still indicates how the writer thinks the ornithologist should be perceived. Thus, concepts of the worldview and social relationships of "an ornithologist" are still valid and for the purposes of this study can be inferred from Chapman's own comments and from his obituaries.

[8] Increasing sentence length accompanying the increasing complexity of scientific discourse would seem to be an obvious delineating characteristic. However, the proposed models for natural history and experimental science discourse do not mention this feature because sentence lengths of sampled authors for both discourse models were uniformly the same. This seeming inconsistency may be explained by the choice of Thoreau and Burroughs, both meticulous writers whose literary backgrounds may have selected for more lengthy expression.

[9] It would be reasonable to assume the reverse in actual practice; that is, the less specialized the discourse, the more definition needed. However, the study found the reverse to be true, likely due to the choice of DNA specialists as experimental scientists. Their extremely specialized, somewhat controversial, DNA articles may have required more extensive use of definition by these authors.

Chapter 5

[1] The labeling of the graphics determined the method of counting occurrences of a particular data display. That is, if three photographs were given the title "Figure 1," then this was considered one occurrence of a photograph. If each was labeled separately, then each was counted. In those few instances in the early volumes of *The Auk* when data displays were not labeled, each occurrence was counted separately. This method was preferred for two reasons: (a) Of interest is the author's choice of representation, rather than the quantity used by a particular author; (b) Some displays, like those using drawings and sonograms, can incorporate many examples

within a particular figure. For instance, a comparative figure of a bird's song may have 30 or 40 individual sonograms per figure. Counting each of these as a separate occurrence would skew the data such that the graphic would appear to have been chosen more than it actually had.

In addition, note that Y-axis values are not uniform among the three line graphs. Thus, slopes of lines cannot be compared directly from figure to figure. Of interest here, however, is the drastic change in graphical presentation during the third phase, 1940–1970, which this study proposes as the period during which American ornithology professionalized.

[2] Because this is primarily a historical study of the changes in American ornithologists' discourse, rather than a statistical analysis of the data, a simple random sample of articles from *The Auk* would not produce an optimal data base, since it is unlikely that articles selected randomly would be spread evenly throughout the publication of the journal. A systematic sample more closely satisfies the needs of this study because such a sample facilitates the collection of articles uniformly across the entire population of interest, thereby providing a cross section of data for each chronological period. Briefly explained, for each stratum of the sample, a unit at random is selected from the first k units, continuing with every kth unit through the rest of the strata. Since selection of each sample is similar to selecting one large sampling unit $[y_i \ldots y_{(n-1)k+i}]$ randomly, a systematic sample may be considered a simple random sample of one sampling unit from a population of k units.

The population of interest was composed of the first article of each issue beginning with the 1890 volume and continuing for each 10-year interval thereafter (that is, 1900, 1910, etc.). These are the same volumes from which were chosen the 444 articles used in chapter 3 to survey American ornithologists' topics, themes, and methodology. Because centrally located samples may add more precision and because the study seeks chronological change, the first article of each issue at mid-decade (that is, 1895, 1905, etc.) was also included in the population. This method produced a population of N=88 articles. Selection of the first article of the issue increased the probability that the subject matter analyzed was of prime interest to the ornithological community since article placement should indicate the community's more pressing issues and its foremost researchers. In order to restrict the study to the most prominent research and the most noted researchers, the database did not include notes, correspondence, and short communications. In addition, since leading articles in early editions of *The Auk* were obituaries for prominent ornithologists, these biographical articles were also bypassed. In order to sample a broad cross section of writing styles, the representative sample should also contain as many different authors as possible. During the selection process, six authors already part of the database were encountered; in these cases, the subsequent article was chosen. This method insured that the database contain a representative cross section of authors and subject matter, equally spread out over the 100 years of *The Auk*.

A pilot study was conducted after first drawing the number one and then selecting a random sample of articles from the above population (N=88) based on this random draw. Thus, the first issue and every 4th issue thereafter produced a sample of 22 articles. The discourse used for the linguistic analysis of these articles consisted of 1,000-word passages from each of the 22 articles in the sample. The method for selecting 1,000 words from each article is as follows. Passages beginning each of the five major sections (Introduction, Methods, Results, Discussion, and Summary/Conclusions or corresponding section headings) and ending with the first sentence break after 200 words (following Bazerman, 1988, p. 167) were chosen. If there were no headings, the first 1,000 words were selected. The same 32 linguistic features described in the linguistic analyses for the discourse models in chapters 2 and 3 were studied. Readers are referred to the explanation in chapter 2 for the methodology and procedure for analysis.

Because occurrences were unacceptably low for a number of linguistic features, a second draw was made, giving the number three. Thus, a second stratum, every third issue of the database, was added. After coding these additional 22 articles, the data from the new sample (n=44) was used to determine the adequacy of the sample size. Confidence intervals were 95% (E=2) for 17 linguistic features, 90% (E=7) for three features (technical terms, personal pronouns, and substantive verbs), and 80% (E=3) for three additional features (simple sentences, interrupting clauses, and citations). These percentages were judged acceptable for the study. One feature was excluded because of insufficient sample size: number of interrupting words. Eight linguistic features were also excluded from analysis because of insufficient data.

See Appendix B for a listing of the articles used in this analysis.

Chapter 6

[1] George Daniels (1967) proposed three stages in the professionalization process. First is a preemption stage in which the body of knowledge needed to conduct science and support claims becomes too esoteric for the general enthusiast. That is, the fact-gathering stage has ended, and formal education is now required. The second stage, institutionalization, gives rise to professional institutions, associations, and organizations that provide a means for collegiality, forums for making new scientific claims, and a body of peers to authorize these claims. During the final stage of legitimation, scientific institutions must obtain public support and acceptance by justifying their existence to society. That is, they must appeal to general cultural values and show the immediate utility of their research. Though Daniels shows the usefulness of his model for describing the professionalization of American science in general, the model does not work well in describing the past 100 years of American ornithological discourse. In particular, due to the persistence of natural history and its discourse, the end of any preemption stage would have to be delayed until at least the 1960s, long after American ornithological institutions had been formed.

References

Abt, H. A. (1981). Some trends in American astronomical publications. *Publications of the Astronomical Society of the Pacific, 93,* 269–272.

Albin, E. (1738). *Natural history of birds* (Vols. 1–3). London: W. Innys and R. Manby.

Allen, A. A. (1927). *Ornithology laboratory notebook.* Ithaca, NY: Comstock.

Allen, A. A. (1930). *The book of bird life.* New York: D. Van Nostrand.

Allen, E. G. (1951). The history of American ornithology before Audubon. *Transactions of the American Philosophical Society, ns 41* (3), 385–591.

Allen, F. H. (1910, September 22). Thoreau's knowledge of birds. *Nation, 91,* 261.

Allen, J. A. (Ed.). (1886). Notes and news. *Auk, 3,* 142–143.

Allen, J.A. (1890). To what extent is it profitable to recognize geographical forms among North American birds? *Auk, 7,* 1–9.

Allen, J.A. (1900). "Aptosochromatism." *Auk, 17,* 327–336.

Allison, P. D. (1979). Experimental parapsychology as a rejected science. In R. Wallis (Ed.), *On the margins of science: The social construction of rejected knowledge* (Sociological Review Monograph 27, pp. 271–291). Keele, England: University of Keele.

Ashmole, M. J. (1970). Feeding of Western and Semipalmated Sandpipers in Peruvian winter quarters. *Auk, 87,* 131–135.

Audubon, J. J. (1827–1838). *The birds of America* (Elephant folio ed., Vols. 1–4). London: Author.

Audubon, J. J. (1831–1849). *Ornithological biography* (Vol. 5). Edinburgh, Scotland: Adam Black.

Audubon, J. J. (1839). *A synopsis of the birds of North America.* Edinburgh, Scotland: Adam and Charles Black.

Auk. (1894). [Editorial page] Vol. 1. Inside front cover.

Auk. (1930). [Information for contributors] Vol. 47. Inside back cover.

Auk. (1940). [Information for contributors] Vol. 57. Inside back cover.

Auk. (1950). [Information for contributors] Vol. 67. Inside back cover.

Auk. (1977). [Information for contributors to *The Auk*] Vol. 94, 191.

Auk. (1982). [Information for contributors to *The Auk*] Vol. 99, 184–186.

Averill, C. K. (1920). Migration and physical proportions. *Auk, 37,* 572–579.

Avise, J. C., Patton, J. C., & Aquadro, C. F. (1980). Evolutionary genetics of birds I. Relationships among North American thrushes and allies. *Auk, 97,* 135–147.

Baird, S. F. (1853–1856). United States War Department. *Reports of explorations and surveys, to ascertain the most practicable and economical route for a railroad from the Mississippi River to the Pacific Ocean* (Vol. 9). Washington, DC: A. O. P. Nicholson.

Baird, S. F. (1858). *Catalogue of North American birds* (Smithsonian Publication 106). Washington, DC: Smithsonian Institution.

Baird, S. F., Brewer, T., & Ridgway, R. (1875). *History of North American birds*. Boston: Little, Brown and Company.

Bartholomae, D. (1985). Inventing the university. In M. Rose (Ed.), *When a writer can't write: Studies in writer's block and other composing-process problems* (pp. 134–165). New York: Guilford Press.

Barton, B. S. (1799). *Fragments of the natural history of Pennsylvania*. Philadelphia: Way & Groff.

Bartram, W. (1958). *Travels through North and South Carolina, Georgia, East and West Florida*. In F. Harper (Ed.), *The travels of William Bartram: Naturalist's edition*. New Haven, CT: Yale University Press. (Original work published 1791)

Batchelder, C. F. (1974). *An account of the Nuttall Ornithological Club: 1873 to 1919*. In K. B. Sterling (Ed.), *Contributions to the history of American ornithology*. New York: Arno. (Original work published 1937)

Battalio, J. T. (1996). The interplay between narrative, education, and exposition in an emerging science. *Journal of Technical Writing and Communication, 26,* 177–191.

Battalio, J. T. (1998). Introduction. In J. T. Battalio (Ed.), *Essays in the study of scientific discourse: Methods, practice, and pedagogy*. Stamford, CT: Ablex.

Bazerman, C. (1988). *Shaping written knowledge*. Madison, WI: University of Wisconsin Press.

Bechstein, J. M. (1877). *Cage and chamber-birds*. London: G. Bell and Sons.

Belknap, J. (1792). *The history of New Hampshire* (Vols. 1–3). Boston: Author.

Bert, E. (1619). *An approved treatise of hawkes and hawking*. London: Richard Moore.

Bewick, T. (1797–1804). *The history of British birds* (Vols. 1–2). Newcastle, England: Beilby & Bewick.

Bishop, L. B. (1900). Descriptions of three new birds from Alaska. *Auk, 17,* 113–120.

Bizzell, P. (1982). Cognition, convention, and certainty: What we need to know about writing. *Pre/Text, 3,* 213–243.

Blome, R. (1686). Hawking and faulconry. In *The gentlemans recreation* (pp. 27–63). London: n.p.

Bock, W. J. (1960). Secondary articulation of the avian mandible. *Auk, 77,* 19–55.

Bolles, F. (1890). Barred Owls in captivity. *Auk, 7,* 101–114.

Bonaparte, C. L. (1825–1833). *American Ornithology* (Vols. 1–4). Philadelphia: Carey, Lea, & Carey.

Bossu, J. (1771). *Travels through that part of North America formerly called Louisiana*. London: T. Davies.

Bowdish, B. S., & Philipp, P. B. (1910). Bird photographing in the Carolinas. With an annotated list of the birds observed. *Auk, 27,* 305–322.

Brewster, W. (1900). Notes on the breeding habits of the American Golden-Eyed Duck or Whistler (*Clangula clangula americana*). *Auk, 17,* 207–216.

Brewster, W. (1910). Resurrection of the Red-legged Black Duck. *Auk, 27,* 323–333.

Brickell, J. (1737). *The natural history of North-Carolina.* Dublin: James Carson.

Brisson, M. J. (1760). *Ornithologie, ou, méthode contenant la division des oiseaux en ordres, sections, genres, especes & leurs variétés* [Ornithology, or the method for the division of birds into orders, families, genera, species, and their variations] (Vols. 1–6). Paris: Chez, C. J.-B. Bauche.

Brooks, A. (1920). Notes on some American ducks. *Auk, 37,* 353–367.

Browne, T. (1684). *Certain miscellany tracts.* London: Charles Mearne.

Brush, A. H. (1985). Editorial. *Auk, 102,* 903.

Brush, A. H. (1988). Editorial. *Auk, 105,* 803.

Brush, A. H., & Clark, G. A., Jr. (Eds.). (1983). *Perspectives in ornithology: Essays presented for the centennial of the American Ornithologists' Union.* Cambridge, England: Cambridge University Press.

Buffon, G. L. L., Comte de. (1778–1786). *Histoire naturelle des oiseaux* [Natural history of birds] (Vols. 1–10). Paris: Imprimerie royale.

Burroughs, J. (1869, June). Birds'-nests. *The Atlantic Monthly, 23* (no. 140), 701–713.

Burroughs, J. (1885). Bird enemies. *The Century Illustrated Monthly Magazine, 31* (new series 9), 270–274.

Burroughs, J. (1889). *Indoor studies.* Boston: Houghton Mifflin.

Burroughs, J. (1895a). Birds'-nests. In J. Burroughs, *Wake-robin* (pp. 103–137). Boston: Houghton Mifflin. (Original work published in 1871)

Burroughs, J. (1895b). *Wake-robin.* Boston: Houghton Mifflin. (Original work published in 1871)

Burroughs, J. (1922). *The last harvest.* Boston: Houghton Mifflin.

Burroughs, J. (1968). *The writings of John Burroughs* (Riverby ed., Vols. 1–23). New York: Russell and Russell. (Original work published 1886–1914)

Burroughs's "Wake-robin." (1877). [Review of the book *Wake-robin*]. *Bulletin of the Nuttall Ornithological Club, 2,* 48–49.

Campbell, B., & Lack, E. (Eds.). (1985). *A dictionary of birds.* Vermillion, SD: Buteo Books.

Cannell, P. F. (1983). An annotated bibliography of the founding of the American Ornithologists' Union. *American Birds, 37* (4), 355–357.

Carew, R. (1602). *The survey of Cornwall.* London: John Jaggard.

Carter, M. (1990). The idea of expertise: An exploration of cognitive and social dimensions of writing. *CCC, 41,* 265–286.

Cartwright, G. (1792). *A journal of transactions and events during a residence of nearly 16 years on the coast of Labrador* (Vols. 1–3). Newark, NJ: Author.

Carver, J. (1778). *Travels through the interior parts of North-America.* London: J. Walter.

Cassin, J. (1856). *Illustrations of the birds of California.* Philadelphia: J. B. Lippincott.

Catesby, M. (1731–1743). *The natural history of Carolina, Florida, and the Bahama Islands* (Vols. 1–2). London: Author.

Chafe, W. (1986). Evidentiality in English conversation and academic writing. In W. Chafe & J. Nichols, (Eds.), *Evidentiality: The linguistic coding of epistemology.* Norwood, NJ: Ablex Press.

Chapman, F. M. (1895). *Handbook of birds of eastern North America.* New York: Appleton.

Chapman, F. M. (1897). *Bird-Life: A guide to the study of our common birds.* New York: Appleton.

Chapman, F. M. (1914). The roseate spoonbill. *Bird-Lore, 16,* 214–217.

Chapman, F. M. (1915). The more northern species of the genus *Scytalopus* Gould. *Auk, 32,* 406–423.

Chapman, F. M. (1917). The life zones of the Colombian Andes. *Bulletin of the American Museum of Natural History: The Distribution of Bird-life in Colombia—a Contribution to a Biological Survey of South America, 36,* 84–93.

Chapman, F. M. (1918). Notes from a traveler in the tropics: I. Down the coastline to Cuba. *Bird-Lore, 20,* 393–397.

Chapman, F. M. (1920). Description of a proposed new race of the Killdeer from the coast of Peru. *Auk, 37,* 105–108.

Chapman, F. M. (1935). *Autobiography of a bird-lover.* New York: Appleton.

Chapman, F. M. (Ed.). (1899). *Bird-Lore, 1,* 28.

Chapman, F. M., & Palmer, T. S. (Eds.). (1933). *Fifty years' progress in American ornithology, 1883–1933.* Lancaster, PA: American Ornithologists' Union.

Charney, D. (1993). A study in rhetorical reading: How evolutionists read "The Spandrels of San Marco." In J. Selzer (Ed.), *Understanding scientific prose* (pp. 203–231). Madison, WI: University of Wisconsin Press.

Chaucer, G. (1980). *Parlement of foulys.* In *Poetical works: A facsimile of Cambridge University Library MS GG.4.27* (Vol. 3, pp. 448–457). Norman, OK: Pilgrim Books. (Original work published 1375–1385)

Christy, B. H. (1933). Topsell's "Fowles of heauen." *Auk, 50,* 275–283.

Cobb, S. (1920). Midsummer birds in the Catskill Mountains. *Auk, 37,* 46–49.

Cochran, W. G. (1977). *Sampling techniques* (3rd ed.). New York: John Wiley.

Coffey, B. B., Jr. (1960). Late North American spring migrants in Mexico. *Auk, 77,* 288–297.

Cole, L. J. (1910). The tagging of wild birds: Report of progress in 1909. *Auk, 27,* 153–168.

Cooch, F. G., Stirrett, G. M., & Boyer, G. F. (1960). Autumn weights of Blue Geese (*Chen caerulescens*). *Auk, 77,* 460–465.

Coues, E. (1871). Progress in American ornithology. *American Naturalist, 5,* 364–373.

Coues, E. (1872). *Key to North American birds* (1st ed.). Boston: Dana Estes.

Coues, E. (1874). *Field ornithology, comprising a manual of instruction for procuring, preparing and preserving birds, and a check list of North American birds.* Salem, MA: Naturalists' Agency.

Coues, E. (1899). The finishing stroke to Bartram. *Auk, 16,* 83–84.

Coues, E. (1903). *Key to North American birds* (5th ed., Vol. 1). Boston: Dana Estes.

Cram, E. B. (1930). Birds as a factor in the control of a stomach worm of swine. *Auk, 47,* 380–384.

Crane, D. (1972). *Invisible colleges: Diffusion of knowledge in scientific communities.* Chicago: University of Chicago Press.

Cruickshank, H. (1964). *Thoreau on birds.* New York: McGraw-Hill.

Danforth, S. T. (1930). Notes on the birds of St. Martin and St. Eustatius. *Auk, 47,* 44–47.
Daniels, G. H. (1967). The process of professionalization in American science: The emergent period, 1820–1860. *Isis, 58,* 151–166.
Davis, D. E. (1950). The growth of starling, *Sturnus vulgaris*, populations. *Auk, 67,* 460–465.
Davis, D. E. (1960). The spread of the Cattle Egret in the United States. *Auk, 77,* 421–424.
De Kay, J. (1842–1844). *Zoology of New York.* Albany, NY: Carroll and Cook.
DeBenedictis, P. A. (1988a). Gleanings from the technical literature. *Birding, 20,* 323–331.
DeBenedictis, P. A. (1988b). Gleanings from the technical literature. *Birding, 20,* 391–401.
DeCrosta, T. (1981). How gardening heals. *Organic Gardening, 104,* 70–79.
Eckstorm, F. H. (1908). Thoreau's *Maine Woods. Atlantic, 102,* 242–250.
Edwards, G. (1743–1751). *A natural history of birds. Most of which have not been figur'd or describ'd* (Vols. 1–4). London: Author.
Edwards, G. (1764). *Gleanings of natural history.* London: Author.
Ehinger, C. E. (1930). Some studies of the American Dipper or Water Ouzel. *Auk, 47,* 487–498.
Eifrig, G. (1910). A winter of rare birds at Ottawa, Ontario. *Auk, 27,* 53–59.
Eiserer, L. A. (1980). Effects of grass length and mowing on foraging behavior of the American Robin (*Turdus migratorius*). *Auk, 97,* 576–580.
Fahnestock, J. (1993). Accommodating science: The rhetorical life of scientific facts. In M. W. McRae (Ed.), *The literature of science: Perspectives on popular scientific writing* (pp. 17–36). Athens, GA: University of Georgia Press.
Farber, P. L. (1982). *The emergence of ornithology as a scientific discipline: 1760–1850.* In *Studies in the history of modern science* (Vol. 12). Dordrecht, Holland: D. Reidel.
Fieth, J., & Male, M. (1992). A celebration of birds with Roger Tory Peterson (J. Fieth & M. Male, Directors). In F. Kaufman (Executive Producer), *Nature.* New York: Public Broadcasting System.
Figgins, J. D. (1920). The status of the subspecific races of *Branta canadensis. Auk, 37,* 94–102.
Fisher, H. I. (1975). The relationship between deferred breeding and mortality in the Laysan Albatross. *Auk, 92,* 433–441.
Fisher, W. K. (1905). Editorial notes. *Condor, 7,* 31–32.
Flower, L. (1989). Cognition, context, and theory building. *CCC, 40,* 282–311.
Flower, L., & Hayes, J. R. (1980). The cognition of discovery: Defining a rhetorical problem. *CCC, 31,* 21–32.
Foerster, N. (1923). *Nature in American literature: Studies in the modern view of nature.* New York: Macmillan.
Fowler, S. P. (1974). Ornithology of the United States: Its past and present history. In K. B. Sterling (Ed.), *Contributions to the history of American ornithology* (pp. 327–334). New York: Arno. (Original work published 1856)
Gardner, L. L. (1930). On the body temperatures of nestling altricial birds. *Auk, 47,* 367–379.

Gibbons, F., & Strom, D. (1988). *Neighbors to the birds: A history of birdwatching in America*. New York: Norton.

Gieryn, T. F., Bevins, G. M., & Zehr, S. C. (1985). Professionalization of American scientists: Public science in the creation/evolution trials. *American Sociological Review, 50,* 392–409.

Goetzmann, W. H. (1959). *Army exploration in the American west: 1803–1863*. New Haven, CT: Yale University Press.

Gould, S. J. (1993). Fulfilling the spandrels of world and mind. In J. Selzer (Ed.), *Understanding scientific prose* (pp. 310–336). Madison, WI: University of Wisconsin Press.

Gradwohl, J., & Greenberg, R. (1980). The formation of Antwren flocks on Barro Colorado Island, Panamá. *Auk, 97,* 385–395.

Grinnell, J. (1920). Sequestration notes. *Auk, 37,* 84–88.

Griscom, L. (1949). *The birds of Concord*. Cambridge, MA: Harvard University Press.

Gross, A. G. (1990). *The rhetoric of science*. Cambridge, MA: Harvard University Press.

Gutting, G. (Ed.). (1980). Introduction. *Paradigms and revolutions: Appraisals and applications of Thomas Kuhn's philosophy of science* (pp. 1–21). South Bend, IN: University of Notre Dame Press.

Halliday, M. A. K. (1993a). On the language of physical science. In M. A. K. Halliday & J. R. Martin (Eds.), *Writing science: Literacy and discursive power* (pp. 54–68). Pittsburgh: University of Pittsburgh Press.

Halliday, M. A. K. (1993b). Some grammatical problems in scientific English. In M. A. K. Halliday & J. R. Martin (Eds.), *Writing science: Literacy and discursive power* (pp. 69–85). Pittsburgh, PA: University of Pittsburgh Press.

Halliday, M. A. K., & Martin, J. R. (1993). *Writing science: Literacy and discursive power*. Pittsburgh, PA: University of Pittsburgh Press.

Hamerstrom, F. (1994). *My double life: Memoirs of a naturalist*. Madison, WI: University of Wisconsin Press.

Harding, W. (1981). Walden's man of science. *Virginia Quarterly Review, 57,* 45–61.

Harding, W. (1982). *The days of Henry Thoreau: A biography*. Princeton, NJ: Princeton University Press.

Harding, W., & Meyer, M. (1980). *The new Thoreau handbook*. New York: New York University Press.

Harper, F. (1930). A historical sketch of Botteri's Sparrow. *Auk, 47,* 177–185.

Hearne, S. (1795). *A journey from Prince of Wales's fort in Hudson's Bay, to the Northern Ocean*. London: A. Strahan and T. Cadell.

Hieatt, C. B. (1983). Stooping at a simile: Some literary uses of falconry. *Papers on Language and Literature, 19,* 339–360.

Hodge, R., & Kress, G. (1988). *Social semiotics*. Ithaca, NY: Cornell University Press.

Holmes, F. L. (1991). Argument and narrative in scientific writing. In P. Dear (Ed.), *The literary structure of scientific argument: Historical studies* (pp. 164–181). Philadelphia: University of Pennsylvania Press.

Holmfeld, J. D. (1970). From amateurs to professionals in American science: The controversy over the proceedings of an 1853 scientific meeting. *Proceedings of the American Philosophical Society, 114,* 22–36.

Hume, E. E. (1942). *Ornithologists of the United States Army Medical Corps: Thirty-six biographies*. Baltimore: Johns Hopkins Press.

Imlay, G. (1793). *A topographical description of the western territory of North America* (2nd ed.). London: J. Debrett.
Information for contributors to The Auk. (1977). *Auk, 94,* 191–193.
Information for contributors to The Auk. (1982). *Auk, 99,* 184–186.
Jefferson, T. (1788). *Notes on the state of Virginia.* 1st American ed. Philadelphia.
Jespersen, O. (1909–1949). *A modern English grammar on historical principles* (Vols. 1–7). Copenhagen, Denmark: Enjar Munksgaard.
Jespersen, O. (1964). *Essentials of English grammar.* University, AL: University of Alabama Press.
Jones, L. (1905). Some April and May work suggested. *Wilson Bulletin, 17,* 22.
Josselyn, J. (1672). *New-Englands rarities discovered: In birds, beasts, fishes, serpents, and plants of that country.* London: G. Widdowes.
Kalm, P. (1770–1771). *Travels into North America.* London: Author.
Kastner, J. (1986). *A world of watchers.* New York: Knopf.
Kendeigh, S. C. (1940). Factors affecting length of incubation. *Auk, 57,* 499–513.
Kendeigh, S. C., & Baldwin, S. P. (1930). The mechanical recording of the nesting activities of birds. *Auk, 47,* 471–480.
Kerpez, T. A., & Smith, N. S. (1990). Competition between European Starlings and native woodpeckers for nest cavities in saguaros. *Auk, 107,* 367–375.
Kessler, F. W. (1960). Egg temperatures of the Ringed-necked Pheasant obtained with a self-recording potentiometer. *Auk,* 77, 330–336.
Killingsworth, M. J., & Gilbertson, M. K. (1992). *Signs, genres, and communities in technical communication.* Amityville, NY: Baywood.
Killingsworth, M. J., & Palmer, J. S. (1992). *Ecospeak: Rhetoric and environmental politics in America.* Carbondale, IL: Southern Illinois University Press.
Kligerman, J. (Ed). (1988). *The birds of John Burroughs: A great naturalist's meditations and essays on bird watching.* Woodstock, NY: Overlook Press.
Kobbé, W. H. (1900). The Rufous Hummingbirds of Cape Disappointment. *Auk, 17,* 8–15.
Kohler, R. E. (1982). *From medical chemistry to biochemistry: The making of a biomedical discipline.* Cambridge, England: Cambridge University Press.
Kuhn, T. S. (1970). *The structure of scientific revolutions* (2nd ed.). Chicago: University of Chicago Press.
Kuklick, H. (1980). Boundary maintenance in American sociology: Limitations to academic professionalization. *Journal of the History of the Behavioral Sciences, 16,* 201–219.
Lakatos, I. (1978). *The methodology of scientific research programs.* London: Cambridge University Press.
Lakoff, G., & Johnson, M. (1980). *Metaphors we live by.* Chicago: University of Chicago Press.
Lankford, J. (1981). Amateurs versus professionals: The controversy over telescope size in late Victorian science. *Isis, 72,* 11–28.
Larson, M. S. (1977). *The rise of professionalism: A sociological analysis.* Berkeley, CA: University of California Press.
Latham, J. (1781–1785). *A general synopsis of birds* (Vols. 1–3). London: B. White.
Latham, J. (1821–1828). *A general history of birds* (Vols. 1–11). Winchester, England: Jacob and Johnson.

Latham, S. (1615). *Lathams falconry or the faulcons lure, and cure: in two bookes*. London: R. Jackson.

Latham, S. (1618). *Lathams new and second booke of falconry*. London: Roger Jackson.

Latour, B. (1987). *Science in action: How to follow scientists and engineers through society*. Cambridge, MA: Harvard University Press.

Lawson, J. (1714). *The history of Carolina; Containing the exact description and natural history of that country*. London: W. Taylor & F. Baker.

Leopold, A. (1987). *A Sand County almanac*. New York: Oxford University Press. (Reprinted from *A Sand County almanac*, by A. Leopold, 1949, New York: Oxford University Press)

Lincoln, F. C. (1930). The menace of oil pollution. *Auk, 47*, 546–550.

Loomis, L. M. (1890). Observations on some of the summer birds of the mountain portions of Pickens County, South Carolina. *Auk, 7*, 30–39.

Lyotard, J. (1984). *The postmodern condition: A report on knowledge* (*Theory and history of literature*, Vol. 10, G. Bennington & B. Massumi, Trans.). Minneapolis, MN: University of Minnesota Press.

Macgillivray, W. (1836). *Descriptions of the rapacious birds of Great Britain*. Edinburgh, Scotland: Maclachlan & Stewart.

Martin, R. F. (1980). Analysis of hybridization between the Hirundinid genera *Hirundo* and *Petrochelidon* in Texas. *Auk, 97*, 148–159.

Mayr, E. (1975). Epilogue: Materials for a history of American ornithology. In E. Stresemann, *Ornithology: From Aristotle to the present* (G. W. Cottrell, Ed., H. J. & C. Epstein, Trans., pp. 365–396). Cambridge, MA: Harvard University Press. (Original work published 1951)

Mayr, E. (1982). *The growth of biological thought: Diversity, evolution, and inheritance*. Cambridge, MA: Harvard University Press.

Mayr, E. (1983). Introduction. In A. H. Brush & G. A. Clark, Jr. (Eds.), *Perspectives in ornithology: Essays presented for the centennial of the American Ornithologists' Union* (pp. 1–21). Cambridge, England: Cambridge University Press.

McAtee, W. L. (1939). Adaptationist naiveté. *Scientific Monthly, 48*, 253–255.

McAtee, W. L. (1974). Economic ornithology. In K. B. Sterling (Ed.), *Contributions to the history of American ornithology. Natural sciences in America* (pp. 111–129). New York: Arno Press. (Reprinted from *Fifty years' progress in American ornithology, 1883–1933*, pp. 111–129, by F. M. Chapman & T. S. Palmer, Eds., 1933, Lancaster, PA: American Ornithologists' Union)

Mearns, B. (1988). *Biographies for birdwatchers: The lives of those commemorated in western Palearctic bird names*. London: Academic.

Merton, R. K. (1973). *The sociology of science: Theoretical and empirical investigations*. Chicago: University of Chicago Press.

Mewaldt, L. R. (1968). Proceedings of the eighty-fifth stated meeting of the American Ornithologists' Union. *Auk, 85*, 101–116.

Morton, T. (1637). *New English Canaan*. Amsterdam, Holland: J. F. Stam.

Murphy, R. C. (1950). Frank Michler Chapman, 1864–1945. *Auk, 67*, 307–315.

Myers, G. (1990). *Writing biology: Texts in the social construction of scientific knowledge*. Madison, WI: University of Wisconsin Press.

Myers G. (1991). Stories and styles in two molecular biology review articles. In C. Bazerman & J. Paradis (Eds.), *Textual dynamics of the professions: Historical and contemporary studies of writing in professional communities* (pp. 45–75). Madison, WI: University of Wisconsin Press.

Myers, J. P., Sallaberry, M., Ortiz, A. E., Castro, G., Gordon, L. M., Maron, J. L., Schick, C. T., Tabilo, E., Antas, P., & Below, T. (1990). Migration routes of New World Sanderlings (*Calidris alba*). *Auk, 107,* 172–180.

Nice, M. M. (1930). A study of a nesting of Black-throated Blue Warblers. *Auk, 47,* 338–345.

Noble, G. K., & Lehrman, D. S. (1940). Egg recognition by the Laughing Gull. *Auk, 57,* 22–43.

Notes and News. (1954). *Auk, 71,* 494.

Notes and News. (1958). *Auk, 75,* 117–118.

Notes and News. (1960). *Auk, 77,* 115–116.

Notes and News. (1961). *Auk, 78,* 669.

Notes and News. (1962). *Auk, 79,* 136–137.

Notes and News. (1963). *Auk, 80,* 225–227.

Notes and News. (1965). *Auk, 82,* 319–321.

Nuttall, T. (1832). *A manual of the ornithology of the United States and of Canada. The land birds.* Cambridge, MA: Hilliard and Brown.

Nuttall, T. (1834). *A manual of the ornithology of the United States and of Canada.* Boston: Hilliard and Gray.

O'Connor, J. G., & Meadows, A. J. (1976). Specialization and professionalization in British geology. *Social Studies of Science, 6,* 77–89.

O'Malley, C. D. (1968). Tudor medicine and biology. *Huntington Library Quarterly, 32,* 1–27.

Palmer, W. (1900). Ecology of the Maryland Yellow-throat, and its relatives. *Auk, 17,* 216–242.

Pennant, T. (1768–1770). *The British zoology.* London: B. White.

Pennant, T. (1773). *Genera of birds.* Edinburgh, Scotland: Balfour and Smellie.

Pennant, T. (1784–1785). *Arctic zoology* (Vol. 2). London: Henry Hughs.

Pera, M. (1994). *The discourses of science.* Chicago: University of Chicago Press.

Perkins, D. N., & Salomon, G. (1989). Are cognitive skills context-bound? *Educational Researcher, 18* (1), 16–25.

Peters, J. L. (1974). Collections of birds in the United States and Canada: Study collections. In K. B. Sterling (Ed.), *Contributions to the history of American ornithology. Natural sciences in America* (pp. 131–141). New York: Arno Press. (Reprinted from *Fifty years' progress in American ornithology, 1883–1933,* pp. 131–141, by F. M. Chapman & T. S. Palmer, Eds., 1933, Lancaster, PA: American Ornithologists' Union)

Petersen, J. J. (1990). Alternative foods of a diet specialist, the Snail Kite. *Auk, 107,* 231–249.

Peterson, R. T. (1984). What are you really? *Bird Watcher's Digest, 63,* 86–93.

Pettingill, O. S. (1992). *My way to ornithology.* Norman, OK: University of Oklahoma Press.

Pettingill, O. S., Jr. (1968). In memoriam: Arthur A. Allen. *Auk, 85,* 192–202.

Physics Survey Committee, National Research Council. (1973). *Physics in perspective. The Interfaces* (Vol. IIB). Washington, DC: National Academy of Sciences.

Piatt, J. F., Lensink, C. J., Butler, W., Kendziorek, M., & Nysewander, D. K. (1990). Immediate impact of the "Exxon Valdez" oil spill on marine birds. *Auk, 107,* 387–397.

Policies on publication costs in *The Auk.* (1978). *Auk, 95,* 219.

Popper, K. R. (1959). *The logic of scientific discovery.* London: Hutchinson.

Porter, J. E. (1986). Intertextuality and the discourse community. *Rhetoric Review, 5,* 34–47.

Potter, E. F. (1970). Anting in wild birds, its frequency and probable purpose. *Auk, 87,* 692–713.

Prelli, L. J. (1989). *A rhetoric of science: Inventing scientific discourse.* Columbia, SC: University of South Carolina Press.

Quirk, R., Greenbaum, S., Leech, G., & Vartvik, J. (1985). *A comprehensive grammar of the English language.* London: Longman.

Reingold, N. (1976). Definitions and speculations: The professionalization of science in America in the nineteenth century. In A. Oleson & S. C. Brown (Eds.), *The pursuit of knowledge in the early American republic* (pp. 33–69). Baltimore: John Hopkins University Press.

Renehan, E. J., Jr. (1992). *John Burroughs: An American naturalist.* Post Mills, VT: Chelsea Green.

Report of the research committee for 1959. (1960). *Auk, 77,* 78–79.

Report of the research committee for 1960. (1961). *Auk, 78,* 242–243.

Report of the research committee for 1961. (1962). *Auk, 79,* 263–264.

Savile, D. B. O. (1950). The flight mechanism of swifts and hummingbirds. *Auk, 67,* 499–504.

Science. (1910, August 11). Review of *Notes on New England birds* by Henry D. Thoreau (F. H. Allen, Ed.). *Nation, 91,* 129.

Sclater, P. L. (1857). Notes on the birds in the museum of the Academy of Natural Sciences of Philadelphia, and other collections in the United States of America. *Proceedings of the Zoological Society of London (Part 25),* 1–8.

Shapin, S. (1982). History of science and its sociological reconstructions. *History of Science, 20,* 157–211.

Shaver, J. M., & Walker, G. (1930). A preliminary study of the effects of temperature on the time of ending of the evening song of the Mockingbird. *Auk, 47,* 385–396.

Sherman, A. R. (1930). The old ornithology and the new. *Wilson Bulletin, 42,* 3–10.

Sherwood, M. P. (1962). Fanny Eckstorm's bias. *Massachusetts Review, 4,* 139–147.

Sibley, C. G. (1955). Ornithology. In E. B. Babcock, J. W. Durham, & G. S. Myers (Eds.), *A century of progress in the natural sciences: 1853–1953* (pp. 629–659). San Francisco: California Academy of Sciences.

Sibley, C. G., & Ahlquist, J. E. (1986). Reconstructing bird phylogeny by comparing DNA's. *Scientific American, 254* (2), 82–92.

Skelton, J. (1983). *The boke of Phyllyp Sparowe.* In J. Scattergood (Ed.), *John Skelton: The complete English poems* (pp. 71–106). New Haven, CT: Yale University Press. (Original work published 1508)

Skutch, A. F. (1971). *A naturalist in Costa Rica.* Gainesville, FL: University of Florida Press.

Slessers, M. (1970). Bathing behavior of land birds. *Auk, 87,* 91–99.

Smith, J. (1612). *A map of Virginia with a description of the country, the commodities, people, government and religion.* Oxford, England: J. Barnes.

Snow, D. (Ed.). (1992). *Birds, discovery and conservation: 100 years of the Bulletin of the British Ornithologists' Club.* East Sussex, England: Helm Information.

St. John de Crevecoeur, H. (1782). *Letters of an American farmer.* Dublin: John Exshaw.

Stanley, E. (1835). *A familiar history of birds* (Vols. 1–2). London: J. W. Parker.

Stevens, O. A. (1936). The first descriptions of North American birds. *Wilson Bulletin, 48,* 203–215.

Stone, W. (1920). Notes and news. *Auk, 37,* 182–188.

Stone, W. (1921). Notes and news. *Auk, 38,* 317–318.

Stresemann, E. (1975). *Ornithology: From Aristotle to the present* (G. W. Cottrell, Ed., H. J. & C. Epstein, Trans.). Cambridge, MA: Harvard University Press. (Original work published 1951)

Suggestions to contributors. (1963). *Auk, 80,* inside back cover.

Sutton, G. M. (1980). *Bird student: An autobiography.* Austin, TX: University of Texas Press.

Swarth, H. S. (1920). The subspecies of *Branta canadensis* (Linnaeus). *Auk, 37,* 268–272.

Taylor, C. A. (1996). *Defining science: A rhetoric of demarcation.* Madison, WI: University of Wisconsin Press.

Tebeaux, E. (1991a). Ramus, visual rhetoric, and the emergence of page design in medical writing of the English Renaissance: Tracking the evolution of readable documents. *Written Communication, 8,* 411–445.

Tebeaux, E. (1991b). Visual language: The development of format and page design in English Renaissance technical writing. *Journal of Business and Technical Communication, 5,* 246–274.

Tebeaux, E. (1997). *The emergence of a tradition: Technical writing in the English Renaissance, 1475–1640.* Amityville, NY: Baywood.

Tebeaux, E., & Killingsworth, M. J. (1992). Expanding and redirecting historical research in technical writing: In search of our past. *Technical Communication Quarterly, 1,* 5–32.

Tegelström, H., Gelter, H. P., & Jaarola, M. (1990). Variation in Pied Flycatcher (*Ficedula hypoleuca*) mitochondrial DNA. *Auk, 107,* 730–736.

Thoreau, H. D. (1906a). *The journal of Henry D. Thoreau* [Walden ed.] (B. Torrey & F. H. Allen, Eds., Vols. 1–14). Boston: Houghton Mifflin.

Thoreau, H. D. (1906b). *The writings of Henry David Thoreau* [Walden ed.] (Vols. 1–20). Boston: Houghton Mifflin.

Thoreau, H. D. (1910). *Notes on New England birds* (F. H. Allen, Ed.). Boston: Houghton Mifflin.

Thoreau, H. D. (1980a). Natural history of Massachusetts. In H. D. Thoreau, *The natural history essays* (pp. 1–29). Salt Lake City, UT: Peregrine Smith Books. (Original work published 1842)

Thoreau, H. D. (1980b). The succession of forest trees. In H. D. Thoreau, *The natural history essays* (pp. 72–92). Salt Lake City, UT: Peregrine Smith Books. (Original work published 1860)

Topsell, E. (1972). *Fowles of heauen*. In T. P. Harrison & F. D. Hoeniger (Eds.), *The fowles of heauen or history of birdes*. Austin, TX: University of Texas Press. (Original work published c. 1614)

Torrey, B. (1906). Introduction. *The journal of Henry D. Thoreau* [Walden ed.] (B. Torrey & F. H. Allen, Eds., Vols. 1–14). Boston: Houghton Mifflin.

Townsend, C. W., & Bent, A. C. (1910). Additional notes on the birds of Labrador. *Auk, 27,* 1–18.

Turberville, G. (1575). *The booke of faulconrie or hauking*. London: Thomas Purfoot.

Turner, F. (1978). The Victorian conflict between science and religion: A professional dimension. *Isis, 69,* 356–376.

Verbeek, N. A. M. (1970). Breeding ecology of the Water Pipit. *Auk, 87,* 425–451.

Wade, J. S. (1940). Canby's "Thoreau." *Auk, 57,* 263–264.

Walkinshaw, L. H. (1940). Summer life of the Sora Rail. *Auk, 57,* 153–168.

Walkinshaw, L. H. (1950). The Sandhill Crane in the Bernard W. Baker Sanctuary, Michigan. *Auk, 67,* 38–51.

Wander, P. C. (1976). The rhetoric of science. *Western Journal of Speech Communication, 40,* 226–235.

Weathers, W. W., & Nagy, K. A. (1980). Simultaneous doubly labeled water ($^{3}HH^{18}O$) and time-budgeted estimates of daily energy expenditure in *Phainopepla nitens*. *Auk, 97,* 861–867.

Weimer, W. B. (1977). Science as a rhetorical transaction: Toward a nonjustificational conception of rhetoric. *Philosophy and Rhetoric, 10,* 1–29.

Welker, R. H. (1955). *Birds and men: American birds in science, art, literature, and conservation, 1800–1900*. Cambridge, MA: Belknap Press-Harvard University Press.

West, G. C. (1960). Seasonal variation in the energy balance of the Tree Sparrow in relation to migration. *Auk, 77,* 306–329.

Wetmore, A. (1920). Observations on the habits of birds at Lake Burford, New Mexico. *Auk, 37,* 221–247.

White, G. (1789). *The natural history and antiquities of Selborne*. London: T. Bensley.

Whitley, R. (1977). Changes in the social and intellectual organisation of the sciences: Professionalisation and the arithmetic ideal. In E. Mendelsohn, P. Weingart, & R. Whitley (Eds.), *The social production of scientific knowledge* (143–169). Dordrecht, Holland: D. Reidel.

Widmann, O. (1895). A winter robin roost in Missouri, and other ornithological notes. *Auk, 12,* 1–11.

Wiens, John A. (1978). Editor's comments. *Auk, 95,* 219.

Wignell, P., Martin, J. R., & Eggins S. (1993). The discourse of geography: Ordering and explaining the experiential world. In M. A. K. Halliday & J. R. Martin (Eds.), *Writing science: Literacy and discursive power* (pp. 136–165). Pittsburgh, PA: University of Pittsburgh Press.

Williams, J. M. (1994). *Style: Ten lessons in clarity and grace* (4th ed.). New York: HarperCollins.

Willughby, F. (1678). *The ornithology of Francis Willughby* (Vols. 1–3, J. Ray, Trans.). London: John Martyn.

Wilson, A. (1808–1814). *American ornithology or the natural history of birds in the United States* (Vols. 1–9). Philadelphia: Bradford and Inskeep.

Wilson, A. (1840). *Wilson's American Ornithology*. Ed. T. M. Brewer. Boston: Otis & Broaders. (Original work published 1808–1814)

Wolfson, A. (1954). The seventy-first stated meeting of the American Ornithologists' Union. *Auk, 71*, 74–84.

Wood, C. A., & Fyfe, F. M. (Eds. and Trans.). (1943). *The art of falconry*. Stanford University, CA: Stanford University Press.

Worth, C. B. (1940). Egg volumes and incubation periods. *Auk, 57*, 44–60.

Yarrell, W. (1839–1843). *A history of British birds* (Vols. 1–3). London: J. Van Voorst.

Ziman, J. (1981). *Puzzles, problems, and enigmas: Occasional pieces on the human aspects of science*. Cambridge, England: Cambridge University Press.

Ziman, J. M. (1968). *Public knowledge: An essay concerning the social dimension of science*. London: Cambridge University Press.

Ziman, J. M. (1984). *An introduction to science studies: The philosophical and social aspects of science and technology*. Cambridge, England: Cambridge University Press.

Zuckerman, H., & Merton, R. K. (1973). Institutionalized patterns of evaluation in science. In N. W. Storer (Ed.), *The sociology of science: Theoretical and empirical investigations* (pp. 460–496). Chicago: University of Chicago Press.

Author Index

Subject Index

9 781567 503951

Printed by Libri Plureos GmbH in Hamburg,
Germany